Change Your Story, Change Your Life

Stephanie S. Tolan

ISBN: 1466214058

ISBN-13: 9781466214057

Dedication

With gratitude to the Author of the Saga

Acknowledgements

Many thanks to the "Spirit Sisters" who gave me feedback (and some arguments), to Bob who put up with and supported the whole project, to RJ and Lisa, for more feedback, Sannie for "working for sushi," and Sharon for creating the StoryHealer website.

Table of Contents

Once Upon a Time...

All we need to do is give up our habit of regarding as real that which is unreal.

–Ramakrishna

The book you are about to read may challenge your view of reality. I hope so. I also hope it will give you a greater sense of the importance of story in human life and the power of imagination that underlies our ability to both tell stories and change them. Once you begin to harness that power, *reality*—that tough, external, rock-solid pattern in which we seem to be caught—can soften and begin to shift.

Let me begin with a bit of what I used to consider my own rock-solid reality. I was a "skinless" child, excruciatingly aware of how it felt to be in this world. On the one hand the scent of crab apple blossoms in the spring, the sound of a bird singing outside my window, the feel of my cat's fur as it purred in my lap, could bring me to tears of appreciation and joy. On the other, I mourned every squirrel killed on the street, felt every taunt and betrayal on the school playground no matter who was the victim, and both imagined and experienced the pain and fear behind every dire story on the news.

That extreme sensitivity did not go away as I got older, but the number of painful or frightening realities I encountered in the world around me and heard about from others began to outnumber and then to overwhelm the moments of joy. Wherever I turned I encountered stories and experiences that darkened my world view. I began to experience depressions, sometimes so deep that I wanted only to escape the pain. I took refuge in books and imagination, but found it harder and harder to contend with *reality* when I emerged.

By the time I was in high school I had become a periodic depressive, sometimes able to push the demons of anxiety and despair away for months at a time. Inevitably, though, they returned. The struggle with those demons, which continued into my adult life, is recorded in page after page of the journals I wrote sporadically over the years, but few people knew of it, because I used every ounce of energy I could muster to keep it hidden. Depression was shameful. I had been raised to believe that fear and pain were signs of weakness. My job in life was to be capable, competent, strong. So that was the image I did my best to project at all times. It was an image designed to both handle and protect me from the *realities* of life in a difficult and dangerous world.

Meantime, I had become a reasonably successful writer of fiction for kids and young adults, readers the age I had been when I first began to contemplate suicide. Remembering so clearly the stresses and difficulties of my own childhood and adolescence, I made sure that all my books included a ray of hope for the reader. I wanted the books I wrote to provide a refuge for other skinless children. The trouble was, the rays of hope I included in my books did little to illuminate my own view of the world. Hope felt little more than an element of fiction, useful in a story, but not reliable, not *real.*

Real was the world, with its wars and inhumanity, its cultures of violence, corruption and greed, its loss of species and habitats, its dwindling resources. When I turned to science, I too often found the *reality* of a random, uncaring universe. The more I focused on what I believed to be real, the darker and more real it became. And so I continued to spend the predawn hours, day after day, struggling with my old, familiar demons.

Today I am no longer a depressive. The demons no longer torment me. My *experience* of life has changed, though the truth of who I am has not. I'm still skinless and I still occasionally catch a glimpse of anxiety and despair lurking in the shadows. What has changed is my view of reality and the way I interact with it. What is real for me now is consciousness, the foundation of all experience. Imagination and story have escaped the pages of fiction and taken their rightful place in my world. *What has changed is the story I tell.*

My own step by step journey to the creation of a new story is not the subject of this book. This book is designed to help you make whatever journey you need or wish to make in order to change your story and improve your own experience of life. It isn't science, philosophy or metaphysics, though it is informed by all of these. Its ideas are both old and new. They've been around in various shapes and configurations for about as long as humans have, and are now finding a toe hold in popular culture and support in leading edge science.

This book is, all in all, metaphor. Story. *My story about story.*

I have written it because I hope to share a way out of whatever rigid reality you might feel yourself caught in. I choose my particular slant on these ideas and the metaphor of story because I am a storyteller. I'm not an expert, a master, a guru. I often have to work at getting free of old stories and creating new ones that serve me better. But I've spent my life creating stories, so I have some experience to share.

This book asks you to consider that you, too, have spent your life dealing with and creating stories, though you might not have noticed. All along, as you have been buying into the stories others tell or making up your own, you've thought of them as real, existing independently of you and your consciousness. This book asks you first to think of them as stories and then to let go of any that aren't working well for you. It suggests that you determine the value of any story not by whether it is true in the moment, and not by how many other people are telling it, but by how it makes you feel as you tell it and how it will make you feel as it plays out in your future. Because our stories *do* play out as we tell them, whether we know that ahead of time or not.

And that is the radical idea at the heart of this book: *that everything we know as reality began as story and is maintained by story. Therefore, like story, it can be changed.*

The principles described in this book are principles for empowerment and they're deceptively simple. Putting them to work can bring a level of well-being into your life that you have never experienced, perhaps never even imagined. But it takes a willingness to venture into new territory and to leave some things behind.

Some of what you leave behind may have become as comfortable as old jeans, as comforting as a fuzzy bathrobe. Some will be old sorrows, pains and difficulties. But even those may have grown comfortable, or at least familiar, over time. It can be scary to let them go. As with every other change in life, as you consider changing your story fear is likely to be your worst enemy. This book intends to help with the fear, too.

It isn't a book to read through and put on the shelf. In fact, it is better not to read it all at once. My suggestion is that you read a bit, try putting that bit to work in your life (there are sections all the way through to help you do that) and then, once you begin to feel and see the effects, read a bit more and go a little deeper.

Human beings are brilliant storytellers. It could be what we do best. You can do this. *If you choose to,* you can bring more joy, more love, more adventure and freedom into your experience through the stories you tell. Both the bad news and the good news is that it's all up to you.

1
My Journey to Story Principle. . .

Out of nothing I have created another,
new universe.

–Janos Bolyai, Hungarian Mathematician

Ever since I can remember I have been a storyteller. In the early years I *pretended* stories, supplementing my own imagination with characters and plotlines borrowed from books and movies. I became a pirate, a cowboy, an Indian brave, Robin Hood, a soldier, a spy, a magician or (on those rare occasions when my character's gender matched my own) a princess. My stories began as solitary imaginings, but whenever possible I commandeered my best friend, other neighbor children, or the occasional visiting cousin, to play the roles of the other characters the story required.

The Written Word

In the fourth grade, my imaginative life took on a new dimension. My teacher, Miss Schultz, assigned us to *write* a story. Reading had always seemed entirely magical to me—the way black marks on paper could become whole worlds filled with sights and sounds and smells, with real people and animals I could care about. But never before had I considered the possibility that I could make such magic myself.

From the moment I finished copying out that first story on lined notebook paper my understanding of the power of story changed. Now my story could be exported to someone else's imagination, someone I might not even know. Something that had existed solely in my own mind could go out into the world all by itself and have an effect on someone else! It was a heady moment, and I decided then and there that I was going to be a writer when I grew up.

It was also that year that I got a memorable lesson about the line between story and reality. We had moved from Ohio to Wisconsin during the summer, so my first day in fourth grade I faced twenty some children I had never seen before. My new plaid dress with matching bows for my braids didn't help me cope with being the one and only new kid in class. When Miss Schultz asked me to tell the class something about myself, it was the storyteller who took over.

I told of my school in Ohio burning to the ground the previous spring. As I spoke the images played out in my mind. There was smoke in the night, a fire alarm blaring unheard, a red glow in the sky, then sirens echoing through the sleeping town as the firetrucks rushed in—too late to save the building. All of us kids, I told my spellbound classmates—and our teachers, too—got a whole extra *month* of summer vacation. It was a good story. Not a *real* one, but a good one. It turned me from new kid to celebrity in an instant.

Lying in bed that night, remembering the undeniable success of my first day at Durkee Elementary, it came to me that I was the only person in the world who would call what I'd told that day a story. Everybody else would call it a lie. My heart clenched in terror. What had I been thinking of? Lies were *verboten* in my family, the very worst crime a child could commit. What would happen when my parents found out what I had done? That they would find out was certain. It was too big a thing to pass unnoticed. Miss Schultz would no doubt call to say how terrible the burning of the school must have been. At the very least she'd mention it to my mother in a parent-teacher conference. I lay awake half the night in an agony of dread.

During that whole year no phone call, no parent-teacher conference unearthed what I had come to think of as The Big Lie. But I couldn't stop worrying about it. Not till I went off to junior high school three years later did I dare to trust that I wouldn't be found out. The line between story and reality needed to be firmly drawn and carefully observed, I decided. Everything belonged on one side or the other.

Reality vs. Story

I did not quit inventing stories, of course. I had become a writer at age nine, and writing eventually became my life's work. For a few years I wrote poetry. I wrote and continue to write the occasional play. Mostly, however, I've made my living writing books for children and young adults.

For the first twenty years my novels were classified as realistic fiction. Except for one small book for early readers (*Marcy Hooper and the Greatest Treasure in the World*), I did not dabble in fantasy. Even then I couldn't shake my preference for reality. One critic complained that the dragon was unconvincingly brought into an altogether realistic story about a regular little girl.

Meanwhile, when my son turned out to be too bright to be readily educated in any available school, I began to write a little nonfiction (first as co-author of *Guiding the Gifted Child*) about the needs of super intelligent kids and adults. Though this writing was nonfiction, therefore *real,* story remained an important part of it for me. It was the drama of true stories that gave that writing its power. I set out to show readers the conflict, the psychological and emotional ramifications, of a child's being out of sync with the expectations of the culture. I wanted to help change the experience that real children were having of not belonging in their world.

As that writing became known, I became a consultant to families and schools about the needs of super bright kids. The more such children I met, the less clear became the line between story and reality. Much of what these kids could do with their intellects seemed impossible, something out of myth or fairy tale. Six month old infants were not supposed to be able to talk; two year olds were not supposed to be able to read two languages; eight year olds were not supposed to be able to ace the S.A.T., understand calculus, or beat expert adult players at chess. But real kids were doing these things, and real families were trying to figure out how to educate them within a system that disallowed their existence.

Then, in the early nineties, a group of children who had been identified as profoundly intellectually gifted, were brought together, along with their parents, for a "Learning Festival," at a ski lodge in upstate New York. A week of activities had been planned that would be appropriate for their extremely unusual minds, but the primary purpose was to let them interact with other kids like themselves—to assure them, perhaps, of their own reality. Those of us deemed "experts" about such children, though we participated, were not certain this gathering would work as it was intended. The kids, ranging in age from five to twelve, were alike in terms of their extreme difference from the norm, but they were wildly different from one another. What would a poetry prodigy, we wondered, have in common with a math prodigy? These children's idiosyncratic passions and intensities seemed too varied to give them a basis for strong personal connection.

We soon discovered how misguided our doubts had been. The ski lodge where the festival was held was almost immediately dubbed "Home Planet," as both the children and their parents experienced a sense of belonging and connection they had never felt before.

I had been working with families of brilliant kids long enough by that time to have come to accept seemingly impossible abilities as normal for this rare population. Thanks to my experience and theirs, my story about reality had changed. But I still maintained a world view informed by the predominant rational/material scientific model of the universe. I did not know, when I agreed to do a writing workshop with the older kids in the group, that my reality was about to change again.

During that workshop the children began creating a group novel. With considerable discussion and a bit of arguing, they came up with a first scene. Then I asked them to sit silently as I made notes so that we wouldn't forget anything. When I finished writing and asked what should happen next, they began to take turns filling me in on the next part of the story. The hair on my arms and the back of my neck rose. They were telling it as if it were a movie they had just seen. When I asked how they all knew the same story, they looked at me uncomprehendingly. "It's what we worked out," one of them said.

"How did you work it out?"

"The same way we worked out the first scene."

"But you didn't *talk,*" I told them.

They were surprised. Whatever their process had been, the working out of the story had felt so natural to them that they simply assumed they had done it the normal, ordinary way.

How *had* they done it? I didn't know. I still don't. Telepathy, or some kind of interaction between minds without words didn't fit my view of reality. But I couldn't deny my (and their) experience. That day was a turning point in both my life and my work. My novel *Welcome to the Ark,* the first book of what the publisher would label a science fiction trilogy (now such works are often called *speculative fiction*), grew out of my interactions with these children.

Here's an important point—it is labeled science fiction (or speculative fiction) because the predominant version of reality operating in this culture at this time is different from the reality these kids actually experience. It's true that *Welcome to the Ark* is fiction. But every unusual mental capacity in that novel, including joining consciousness not only with other humans, but with trees or rocks or clouds, is a mental capacity that humans can, and some do, experience. How are we to determine what is and is not real?

In her chapter in Erwin Laszlo's book, *Science and the Reenchantment of the Cosmos,* evolution biologist Elisabet Sahtouris deals with this question, first giving a definition of reality from *Webster's Online Dictionary*: "something that is neither derivative nor dependent but exists necessarily," and then pointing out that "The only thing fitting this definition is direct experience, for once any experience is reported to another, whether by a three-year-old, a scientist, or a theologican, it clearly becomes derivative." [1]

This doesn't mean that something we personally haven't experienced can't be real, but that something we *have* experienced "exists necessarily" for *us*, whether anyone else has experienced it or not.

The Real Question

Changing where I drew the line between story and reality allowed me finally to begin to question whether it can ever be drawn with permanence or certainty. We live our lives in a material world we can feel and see and smell and taste and touch. We experience that world as real. But at the same time we are making a personal interpretation of all of those sensory experiences, telling ourselves a story about them. Which is the *real* bite of liver—the one I say is delicious with onions and mashed potatoes, or the one my husband calls disgusting? Which is the *real* story about the planet we live on—the one scientists tell us today (assuming they could come to agreement on it) or the one that will be discovered tomorrow?

Today's scientific equipment gives us a world view vastly different from the world view of people living on this same planet a thousand years ago. But is ours *more real?* When people believed the earth to be flat, their experience of it was in agreement with their belief. It looked flat, it felt flat, it made logical sense that for them to be standing upright on it, it pretty much had to *be* flat. Nothing in their lives challenged their version of reality.

The idea of a spherical earth remained for most humans just a story, even after evidence had been gathered by a few, until people actually began to move around the world on ships without falling off. Only when the spherical earth began to fit both what people were told and their actual experience could that story become real to them. By the time we had that first photograph of the "blue marble" of our planet taken from space, it was the flat earth that had become no more than story. If we stand in a Kansas wheat field today the earth may appear flat, but we have learned a different interpretation of that sensory information. Telling a new story, we live a new reality.

When I was a child and radios were furniture, Dick Tracy's wrist radio was only story. *Destination Moon* was a science fiction novel I loved when I was ten. Less than twenty years later, I got chills as I listened to the first message sent back to earth from Neil Armstrong as he stepped onto the moon's surface. The moon as human destination had become reality. Twenty-five years ago the idea of faxes, pda's and cell phones (cell phones that can also make movies and send them over the air-waves!) would have been no more than story to me, no matter how many scientists understood that such inventions might one day be real.

The first expressions of new technology often come into public perception through science fiction, written by people who use their imaginations to venture out of our current consensus reality and tell a new story. Visionary scientists, of course, must also take the first steps into a new reality in their imaginations, as Einstein so often said.

Mind Stuff

When I visit schools I often ask kids whether they have ever felt sad when they finish reading a book because the characters are gone from their lives. Almost all of them agree that they have. We then talk about how real the story has felt to them, how alive the characters seemed, and how they wish the story could continue so that they wouldn't lose the characters they have come to know and love.

I ask them where the characters in one of my books were *before* I wrote it. "In your head!" they usually answer.

"What about before I got the idea?"

There is usually a puzzled silence. Then someone says tentatively, "Nowhere?"

"Exactly," I say. "They didn't exist *at all.*" Then I go on to talk a bit about the creative power of imagination—a power they all have and have used for themselves but may not have thought about very much. We talk about where the characters and the story are "right now," and the kids come to realize that they exist only in our minds. They are made, however real they feel, of nothing more than thought, imaginings, "mind stuff."

"Will the characters go on living?" I ask them. "I don't mean will they go on doing new things, but will they always be alive in this story?"

Quite often this conversation leads a child to suggest that as long as anyone can read the book, the characters will go on being alive. Sometimes another child will then say that even if all the books were destroyed, the characters would go on living in the minds of anyone who had read the story.

Once a child said, "That's the way people go on living after they die—in the minds of the people who knew them."

Another added, "That's the way *anything* that happened in the past goes on—just in our minds—in our memories."

The kids go away from our discussion with a new way of thinking about stories and about imagination. They see that imagination—mine first and then theirs—generates, out of nothingness, or from symbols on a page, the characters, the setting, the plot of a story that has felt quite real to them as they read it.

It was conversations like these and a little exploration of quantum physics that gave me what I call Story Principle.

[1]"From a Mechanistic and Competitive to a Reenchanted and Co-Evolving Cosmos," in Laszlo, Ervin. Science and the Reenchantment of the Cosmos, The Rise of the Integral Vision of Reality. Inner Traditions: Rochester, VT. 2006, pp. 107, 108.

2
Story Principle 101

What is life? An illusion, a shadow, a story.
–Pedro Calderon de la Barca

Let's begin with a streamlined overview, a kind of primer of the way Story Principle works.

As we move through our lives, moment by moment, day by day, each of us is telling ourselves a story—about ourselves, about what is happening to us, about what we can or should do about it, and what it all means. Not everyone makes a living creating stories, as I do, but in a very real sense we all make our lives that way. The biggest difference between the stories I create for my novels and the ones we are all creating in our lives, is that I always know I am working with fiction, while most people believe their lives to be Reality.

> ***Of course my life is reality! I can't just change things the way a writer can change the characters and plotline of a story! Reality is solid. Reality is—REAL!***

Consider for a moment the possibility that *everything* is story. (Or—if you aren't ready to go that far—that everything in our subjective experience is story.) The wonderful thing about story, *real* or otherwise, is that because we create it, we can change it—in any given moment.

Here's a personal example: I was to give the talk that was the seed for this book on a Saturday morning at 10:45. Friday evening I developed severe head congestion and a cough that kept me up most of the night. I was staying at a hotel several blocks from the convention center where I was to speak, and knew that there was a CVS pharmacy between my hotel and the center. So I left early to stop on the way and get some sinus medication, hoping to clear up the congestion enough to speak clearly.

At CVS I discovered that the medication I wanted, which contained pseudoephedrine, couldn't be simply taken off the shelf and purchased in Kentucky as it could at the time in North Carolina. All that was there was a card that I could take to the cash register, where I presumed the cashier would get the medication. The store was very busy that morning. I took the card and a bag of cough lozenges, got into the shortest line, and waited.

When I finally reached the cash register I was told that I was in the wrong line for making a purchase—*this* line was for lottery tickets only. So I checked my watch (the extra time I'd allowed was fast disappearing!) and got into the longer line. When the four people in front of me had completed their business, the cashier looked at the card I was holding out to her, and said, "Sorry, you can't have that. You have to get it from the pharmacy. It isn't open on Saturdays."

So far we've been talking about what we call *reality*. Here's where *story* comes in. Fifteen or twenty years ago, back when I was a pessimistic depressive, I would have been telling myself this story: "Of course I can't just buy what I need. Nothing's ever easy." I would have told myself (as I hear people saying at the grocery store all the time), "I *always* choose the wrong line!" And finally, I would have moaned that given my dreadful luck the very fact that I needed those pills was practically a guarantee that I wouldn't be able to get them, and now I was going to be late for my presentation besides. I would have been stressed and miserable, and I would have attributed my misery to the *reality* of my experience. I wouldn't even have suspected that I was telling myself a story about it, much less that the story I was telling could really affect what was happening to me!

But I had discovered Story Principle, which agrees with quantum physics that reality is not as solid as it appears. So I had changed the story I tell myself about who I am and what my place is in the world. My life, my *Real Life*, has changed with my story. Now I say, "I have whatever I need whenever I need it, wherever I need it, for as long as I need it." And it keeps turning out to be true!

But how…?

Hold that very tricky question for a later chapter. This is the primer version, remember. You can begin using Story Principle without knowing the answer.

Instead of getting more stressed with each setback that morning, I resolutely told myself the story that everything was fine and I had all the time I needed. When I left CVS, a full fifteen minutes later than I'd intended, every traffic light, as I came to it, gave me a "Walk" signal. Though I hadn't even been in the convention center yet, I easily found the room (which happened to be right by the entrance I'd used) and arrived at 10:43. I spoke the first words into the microphone exactly when the talk was scheduled to begin.

I told my audience this example of changing my story and explained that since I hadn't been able to buy the medication, I must not need it (because *I have what I need whenever I need it*). My head congestion cleared up as I spoke the first few words and didn't return until hours later!

I suspect most of you have had something annoying happen to you early in the morning and announced to yourself, "It's going to be one of those days!" I further suspect that you then watched your day disintegrate into a parade of aggravations.

You don't have to believe me that changing your story would have changed your day (though I do say it's possible). Next time just try acting *as if* that could be. Catch yourself telling a negative story, change the story, and watch what unfolds in your life. At the very least, you'll feel better about your trying day (not a bad thing). At best, what seemed initially to be a problem could turn out to be the necessary lead-up to a cascade of positive experiences.

You're hurrying to get to a meeting at work that you aren't quite ready for. As you get into your car, you spill your coffee. Instead of cursing, as you might normally do, and stewing about how this could make you late, you note how lucky you are that the coffee didn't spill on your suit or your notes for the meeting. At least you don't have to go back inside and change. As you search for something to sop the coffee up with, you find the fountain pen a co-worker was looking for at the office yesterday after the two of you went to lunch. Thinking that you probably wouldn't have found that pen for weeks if the coffee hadn't spilled, you drive away feeling a bit better. Your drive goes unusually quickly so you have time when you get to work to take the pen to her desk. She happens to be having a phone conversation with the very person you've been trying to track down for days to get some information for this morning's meeting. She gratefully takes her pen and turns the phone over to you. You get everything you needed to know and go on to the meeting, fully prepared.

That's the kind of thing that can happen when we change our habitual negative stories. It takes practice to tune in, to hear what we're telling ourselves, and to begin to think of that as story rather than reality. But the more we try it, the better we become, first at noticing our story, and then at changing it. As we change it, our experience changes. Inevitably. We adults tend to have some trouble with this; most children, with fewer ingrained stories and shorter histories of unpleasant experiences, are able to put this principle to use quickly and easily.

Maybe changing the story could work for minor annoyances. But what about a really bad experience? Just changing what we tell ourselves about it won't make it better. Story Principle can't be useful then.

On the contrary. It is when faced with our toughest experiences that the principle becomes most important and has the most powerful positive effect. Sometimes the stories we tell ourselves as our most difficult experiences unfold—stories about who we are, why these things are happening to us, and what the effects of them will be on the rest of our lives—can make the difference between surviving our difficult times and being defeated by them.

Let's look at one of my own favorite stories, Tolkien's *Lord of the Rings.* It can help us to consider becoming the hero of our own story even in the most harrowing of times. Here we have Frodo the hobbit (small and powerless—hardly heroic) who would prefer to stay in his cozy hole, warming his toes by the fire and eating plenty of good food. But when he is given the task of carrying the Ring of Power to the Cracks of Doom, he sets out, in spite of his preferences and his fears and the likely overwhelming odds he is bound to face. Why? Because he's the hero of the story. And because he believes his task (if not himself) is important and meaningful.

Because he's small and powerless, he expects to need help. That very expectation attracts it to him, and the Fellowship of the Ring is created. Frodo has critical help throughout the story, but he remains the hero—the ring itself must remain with him. No one else can carry it for him.

Sometimes Frodo is extraordinarily frightened, because the Nazgul (the nine flying Ring Wraiths) who are after him are formidable and terrifying, as are the orcs and goblins and all the other obstacles between his cozy home fires and the Cracks of Doom. Sometimes he's hungry and exhausted. He gets hurt along the way. Who could blame him for getting discouraged, even refusing to go on? But he goes on. He's the hero.

Life can feel like Frodo's story sometimes. We are bathed in the fearful "realities" of our culture. What we hear from the media all around us is the message that external forces (and enemies) are always at work to harm us if we are not constantly vigilant against them. Nazgul, we are told and told again, are always hovering above us. The images of our enemies may change, but they're always there.

The more sensitive, perceptive and aware we are, the more likely we are to take in and believe, even exaggerate for ourselves, these cultural warnings. Meantime, we see the evidence of the scary stories being worked out all around us. To doubt them, we feel, would be sheer denial. At the same time, all of us encounter patches of quicksand and pitfalls—even a Nazgul or two—along our journey. If we are telling ourselves the fearful story of our own powerlessness and victimhood, we can be readily overwhelmed by what seem impossible odds.

But if, instead, we identify ourselves as the *heroes* of our story, we can accept the possibility that wherever there is quicksand or pitfall or Nazgul, there is also some way to get past it. Just telling ourselves that there is a way past it allows us to begin looking for that way instead of giving up. Our sense of our own power actually adds to our strength, our ingenuity, our adaptability.

For those who enjoy computer games, a game metaphor may be useful. It's clear that no matter what the character on the screen must face, there is always a tool or weapon available somewhere in the world of the game, if the player can figure out what it is and how to acquire it (and he *can* because that's the whole point of the game!), that will allow the character to go on.

It isn't only the story we tell ourselves in the middle of a difficult experience that counts. There's also the story we tell ourselves afterwards. When something painful has happened, the person who says "I was a hero in this situation, and *because* I was—and am—a hero, it did not defeat me," will have a far better chance of healing whatever wounds she might have received and moving on, than someone who tells herself, "This unfair and awful thing happened and no matter what I do things like that will go on happening to me all the rest of my life, because that's the way the world is."

In our most challenged moments, when something very painful has happened in our story, we can tell ourselves that sometimes we just have to lie down for a while, wrap ourselves in a blanket, and recover before we go on. Even a hero may say, "I'm going to take a break now till I feel better." Afterwards, heroes get up again.

Here's another example of the power of heroic self-perception. Consider Cinderella, who did not have such an important task as destroying the Ring of Power. Like most of us, what she had to handle was her life. If Cinderella had bought the story that her stepmother and stepsisters had all the power, that she was going to be their scullery maid forever, would she even have *wanted* to go to that ball? Hardly! She'd have *known* no prince would look at a scullery maid. She'd have sat by the kitchen fire moaning that she had nothing, "Just these rags and cinders and no chance ever to change my miserable life."

But Cinderella wanted so badly to go to the ball that the intensity of her desire conjured up for her a fairy godmother who could work the miracle of the mice and the pumpkin and the gorgeous dress. This is no *deus ex machina* story where a powerless victim is saved by a magical outside force. If Cinderella had considered herself a victim, would a dress, however magical, have given her the nerve to walk into the ball and dance with the prince? (Back in my days of "ball going," no mere gorgeous dress could have kept me from standing alone and forlorn against the wall—as I did—telling myself that nobody wanted to dance with me.) Cinderella not only knew *somebody* would want to dance with her, she dared to tell herself it could be the prince himself. She was the hero and this was *her* story!

When the clock began to strike twelve and she had to flee the ball, Cinderella didn't say, "See, this kind of thing always happens—just when the prince is getting to like me time runs out and I have to drag myself home with one glass slipper and one bare foot." No, she told herself that she had just had the most wonderful night of her life, and nobody could ever take it away from her.

Later, when the king's men come around with the glass slipper, she *demands* to be allowed to try it on. Voila! Cinderella, with a little help from the allies she summons through the power of her story, creates her own happy ending.

Just as we have our own story, of course, so does everybody else. But we don't have to take their story, or any part of it, for our own. If someone says, "You're nothing but a scullery maid!" (or a jerk, a liar, an idiot—whatever might be said to our faces or behind our backs) we can shrug and say, "That's *your* story, not mine!"

With this vision of ourselves as the hero of our story, we can be better prepared to confront our obstacles, our challenges, our Nazgul, and even our wounds. Then, in our daily lives, if we hear ourselves saying "I am…" followed by a less than heroic attribute like "…sick and tired," we can change that before we find ourselves having to take to our beds.

This Story Principle seems altogether too simple, too "Easy" to work. How am I supposed to be able to believe it?

I'm not suggesting you *believe* it. Certainly not on my say so. I'm suggesting that you try it for yourself. This chapter is called Story Principle 101 because it sketches out the basic principles. The next section provides some ways to begin using them. You can put Story Principle to work and begin changing your life in really powerful ways right away. Try it.

Putting It to Work

The first thing most of us need to do to put Story Principle to work is to discover the stories we are telling ourselves that aren't useful. How do we do that? Begin by listening.

Listen to other people.

It's always easier to notice someone else's stories than our own, since our own feel like realities! You can do this among your friends, your family, your co-workers, and by eavesdropping on strangers at restaurants, in airports, or talking on their cell phones.

Remember—Story Principle says that negative stories bring negative experiences and positive stories bring positive experiences. Positive stories, by the way, are stronger—but more of that in a later chapter.

What you want to find out is how much of the time what people say is a negative interpretation of something that is going on in their lives, or a negative prediction about something that hasn't happened yet. And how often are they positive, about past, present or future events? You'll quickly recognize the difference, sometimes just from the tone of voice.

Unless you hang out with an extraordinarily cheerful and optimistic group of people, you are almost certain to discover that a large percentage of what people share with each other consists of complaints, worries, fears, aggravations, annoyances and predictions of disaster. If negative stories, told over and over, either maintain or bring into existence negative experiences, they are clearly *not helpful.*

Not long ago I was at the airport for a flight to California that was due to leave at 9:40 am. A little before 9:00 there was an announcement from the podium that the flight would be delayed—there was a mechanical problem. "You may leave the boarding area, but be back in half an hour for an update." Considerable groaning greeted this announcement. A number of people loudly prophesied missing their connecting flights and got on their cell phones to rearrange their plans.

Fifteen minutes later the announcement was made that the mechanics had determined that the problem was not fixable. The plane could not fly. They were going to have to search the system for another plane and there was no estimate of time delay. Immediately the waiting passengers, even those who had remained stoic after the earlier announcement, went into crisis mode. Cell phones came out and meetings were canceled. Though clearly no one had any idea how soon the flight would leave, virtually everyone with tickets on that flight predicted that the delay would be hours long and their travel plans for the day would be ruined. I heard not one single passenger predict that the situation might be handled quickly. There was not a positive voice in the place; even the airline personnel were carefully neutral.

As you listen to other people's conversations, be alert for negative predictions, negative interpretations of the motives of others, negative past events being told and retold, or present experiences being mined for their most negative aspects. Get used to recognizing them and take notice of how much negative storytelling surrounds you all the time.

Listen to yourself.

Once you have begun noticing other people's negative storytelling you'll be ready to turn your attention to yourself. You don't have to monitor your every thought. Just listen to what you hear yourself saying.

You may be amazed at how often you predict a problem for yourself before one emerges. "I'll never get this job done in time." "Whenever I get to bed this late the kids wake me up even earlier than usual." "I need to get my car fixed but mechanics always think they can rip a woman off." "My boss wants to see me—I wonder what he's going to criticize me for this time!" "Oh, sure that politician *says* he'll do it differently, but they always lie and nothing ever changes."

Once you have a sense of how often negative statements come out of your mouth, you'll have a sense of the general trend of your thoughts. What you say aloud is only a small representation of the thoughts that lie behind the words. Are you occasionally negative, often negative or mostly negative? Answering this question will give you a sense of how much work lies ahead to begin changing your stories and your experiences.

Experiment with changing a story.

It's a good idea to begin with a small, fairly unimportant one. When you're driving you might catch yourself thinking, *That light up there has been green so long it'll turn red just as I get there.*

You could change it to, *it'll stay green till I get through,* but that may not be the best choice for a newcomer to Story Principle. If you say that and the light turns red, your rational self is likely to ridicule the whole effort. *The timing of traffic lights has nothing whatsoever to do with you!* Depending on how nice you're used to being to yourself, that rational voice could go on a rampage: *who do you think you are to expect traffic lights to work for you? Who made you god[dess] of traffic lights?*

This sort of thing can be hard on a beginner.

A more general story change, such as *I'm going to get where I'm going in perfect right timing* is a better choice. That way, if this one light turns red as you approach, it won't scuttle the new story. You can repeat and refine the new story as you go, suggesting, for instance, that the lights begin going your way and traffic moves smoothly and easily.

Though you can begin telling a positive traffic story any time, the best time to put it in place is before you begin the trip. As you start the car, take a couple of relaxing breaths and say, "This trip moves smoothly, easily, under grace. Time is not an issue." The idea is to defuse whatever stress you may be feeling as you start, because stress has a way of escalating. Even if you've started late and it seems time is woefully short, if you can drive away feeling calm, confident and relaxed you create the likelihood of a calm, relaxed drive and an arrival time that will allow you to maintain that feeling. Starting with a high degree of stress, feeling rushed and frantic, can quickly turn any drive into the sort of meltdown trip where lights turn red at your approach, intersections are jammed, school buses stop at every third corner, and you have to pull over to let a fire truck roar by.

Take notice of how the story *feels*.

Feelings fuel the process of turning stories into experience. The more powerful your feelings about a story, the more quickly and surely it will affect your experience. Suppose you catch yourself telling the story that you don't have time to complete a job you need to do. If that story makes you feel overwhelmed, try to change not just the words, but the feeling behind them. It's not so useful to plaster positive words over strong negative feelings.

There are various strategies to begin changing negative feelings. First acknowledge them (denying them doesn't help a lot), then shift your focus to something that might soften them a bit. Try to remember a time in your life when you accomplished a big job successfully. Remember how it felt. Remind yourself that you actually did it. Try to give the whole process the flavor of a game. This is why it's good to start small, with stories that don't carry a lot of emotional charge.

Here's a personal example: When the flight mentioned above was delayed, I found myself starting to feel the same sort of distress as the other passengers. My plans for that day were not flexible. I was scheduled to speak at a conference only two hours after the plane was due to land, and it was too early to reach the organizers, who were three time zones away. I took some slow deep breaths, reminded myself that, with traffic at the other end, I had probably a little over an hour of wiggle room, and told myself that things would work out okay, one way or another. I've often had flights delayed for very brief times for minor glitches, so I reminded myself of that and went back to reading my book.

But when the second announcement came, I could feel myself joining the other passengers in crisis mode. Negative energy can be very contagious. I found that I couldn't stay in the gate area, surrounded by distraught people who had already decided their day had been ruined, and hold onto a positive story myself.

So I moved to an empty gate area, sat alone, and did some more deep breathing. Once I got calm, I said to myself, "Whatever happens here, everything I need to do will get done and my trip will go smoothly." I didn't try to figure out how that could happen, I just told the story that it would. Again, I went back to my reading to keep my attention focused away from the problem. Less than ten minutes later another announcement was made—a plane had been found in a nearby hangar. As it turned out, the flight departed less than an hour late and made up most of the time in the air. No connecting flights were missed and I can only assume that the other passengers' days went as smoothly, thereafter, as mine did. None of them thanked me, of course, but I grinned to myself as we deplaned, telling myself that my one positive story had trumped all the rest. It certainly sent me off to my speech feeling good!

Get yourself an Easy Button.

You may have seen the Staples commercials that suggest using their Easy Button. Did you know you can buy one? They cost $4.99 and the proceeds support the Boys and Girls Clubs of America. It looks just like the one in their commercials, and when you press it, a cheerful male voice says, "*That* was easy!" There are instructions for its use: "1. Identify a difficult situation. 2. Press your easy button. 3. Listen to its reassuring message. 4. Smile and get on with your day. 5. Repeat as necessary."

This has to have been designed by a person who understands the effect of story. Note that you aren't expected to wait until you've *handled* the difficult situation—you hit that button the moment you've identified the difficulty. And you're not told to get busy, then, and solve it. The *story* the button provides for you is that the solution comes (having been easy) as you, smiling, go about your day.

I keep one within easy reach on both floors of my house to remind me to put Story Principle to work for me. As a survivor of a family with a deep story about the value of hard work (not just work, mind you—*hard* work), I find the Easy Button an important antidote. Otherwise I am likely to make even the easiest task difficult just to convince myself that it's worth doing.

If you can't find an Easy Button, or Staples quits making and selling them, you can make your own—put a red circle on a piece of paper or cardboard, write Easy Button across it, and touch it whenever you identify a difficult situation. Then say to yourself (preferably out loud), "*That* was easy!" I've given so many of these away to family and friends that whether one is handy or not, it's an often heard statement in my world. The great thing about it is that it almost always makes us laugh.

Find an ally or two or three.

Learning to work with Story Principle isn't the sort of world-saving task that Frodo faced in

The Lord of the Rings, but it can really help to have the support of someone else who is doing it, too, a "Fellowship of the Story," whether that means one person willing to take this journey with you or several. Putting Story Principle to work *can* be done alone—it only takes one teller to create a story—but having psychological and emotional support as you do it smoothes the way. This is true of beginners and old timers alike.

The culture we live in is not supportive of the Story Principle view of reality. So each of us who wishes to empower ourselves and enrich our life experience by choosing the stories we tell is like a salmon swimming upstream in the river of the collective consciousness. If we tell people what we're doing we're likely to encounter a barrage of negative responses, from doubt to outright ridicule. It can be hard to stand against that force. But if you can find someone who is open to trying this with you, you can shore each other up. Share the opening chapters of this book with someone who seems open to these ideas and see whether they'd be interested in joining you.

It can be easier to keep a playful focus if you're playing with someone else. You can share your successes, and support each other's stories when they don't seem yet to be working out. It doesn't matter whether you're both beginners or whether you find an ally who is already on the path of creating his or her own reality. The important task of the "Fellowship of the Story" is to hold each other's stories in consciousness, to remind each other of what you're trying to do, and to offer each other a helping hand out of the quicksand or around a pitfall.

Meanwhile, it's a good idea whether you're working alone or with others to keep notes on the stories you manage to change and the results of changing them.

Don't be discouraged if Story Principle doesn't seem to work the first time you try it. Remember to treat it as a game and try again. The more you practice telling positive stories, the better you'll get at stopping the negative ones and changing the pattern of the thinking that created them. If you've learned to play a musical instrument or even to work the controller of a computer game, you know that the early going can seem awkward and slow. But the more you practice, the easier it gets.

Even when you've been practicing, you can sometimes suffer a kind of mental meltdown, which can be a useful learning process as you continue forward. Not long after the successful story-telling experience of that plane trip to California, I was supposed to speak at a conference in Vancouver. The organizers had asked for my passport number so they could arrange for my airline tickets. I got my passport from *the place I always keep it,* took it down to my office and sent the number off by e-mail.

The afternoon of the day before I was to fly, I went to *the place I always keep it* and my passport wasn't there. It had been six weeks since I'd had it in my hand and I didn't remember whether I'd sent the e-mail from my laptop upstairs or from my office. Chaotic and messy as I am, I am very, very careful with my passport, so I couldn't believe I hadn't put it away. I began my search fairly calmly, sure that I'd find it quickly. I didn't.

When my husband came home two hours later, I was frantic. I had looked every single place it could be in the entire house and most of the places I knew it *couldn't* be.

It was two weeks past the date when passports had become requirements for traveling between the U.S. and Canada. My head was filled with catastrophe stories of having to call and cancel my appearance, of going anyway and being kept from returning home. My husband's calm drove me even crazier.

When it became clear that I wasn't going to find the passport, he made some phone calls and got the number for the duty officer at the passport office in Washington

DC. He told me to take a couple of deep breaths, then dialed it and handed me the phone. The officer assured me the requirements were still so new that I could get both into Canada and back into the states if I took my birth certificate and a photo I.D. Once the story of having to cancel and the story of being detained in Vancouver had been deleted, I began to breathe normally again.

Five minutes later I went back to my office and realized that the one and only "impossible" place I hadn't looked was the bottom drawer of the file cabinet next to my computer. I opened that drawer and there was my passport, stuck between two manila folders! The moment I saw it I remembered that because of an injury I'd had, I had been using that open drawer as a footrest when I was asked for my passport number. The passport must have fallen in and—once there—become invisible and forgotten. That was why I hadn't put it in *the place I always keep it.*

It was an important reminder that when we get caught up in negative stories, fueled by intense feeling (particularly around personal trigger issues like my inherent messiness and lack of organization) we cut ourselves off from the very help we need. Once my frantic catastrophizing had stopped, my intuition was finally able to kick in and make itself heard.

Another important part of that failure at Story Principle was my awareness that I was failing—in spite of the fact that I was coming close to finishing the writing of this book!—and my harsh judgment of myself for letting myself freak out, even as I freaked out. It is critical not to start beating up on ourselves when we aren't doing as well as we'd like.

Human beings are complicated creatures, with old stories caught up in tangles of emotions. As similar as the two travel stories might seem from the outside, the critical difference was that when the airline announced the problem with the plane, I had nothing to do with the glitch. The disappearance of the passport, on the other hand, was plainly "my fault." I'd failed to put it away in the first place and it was my chaotic office that was keeping me from finding it. It wasn't until my husband brought his calm, non-judgmental viewpoint into the situation that I could stop berating myself and calm down a bit.

The speed of the turn-around, finding it as soon as it didn't matter so much, shows the power of Story Principle in action. Once I could actually use it, it worked. As you begin practicing, do your best to keep the whole process as light and playful as possible. If it works, celebrate. And write it down somewhere to remind yourself. If it doesn't work, let it go. It isn't a test. Your lack of achievement won't go down on your permanent record. You can always, always, try again. And the more you do it, the easier it gets.

When you've begun to be successful at changing some of your small stories, it's time to read on. "*That* was easy!"

In the next chapter we will deal with the conceptual aspect of Story Principle, the idea and theory behind it. If you want to stick with the practical application of the theory to your life, you may prefer to skip that part and return to it later.

3
Story Principle and Cosmology

In the beginning was the word...
-John 1;1

Story Principle is more than a practical way to improve our daily lives. It's a metaphor for what Douglas Adams, who wrote *The Hitchhiker's Guide to the Galaxy*[1] called the question of "life, the universe and everything." According to Adams, the answer to that question is forty-two. According to Story Principle the answer is Mind, or Consciousness, or Imagination.

The basic idea that I call Story Principle can be found in the mystical aspects of most of the world's religious traditions. It is the foundation of what is known as "New Thought." And since the early decades of the twentieth century quantum physicists have been moving science toward it, one small step at a time. It consists of two primary tenets: Unity underlies all the diversity of the universe; and all that exists in the material realm begins in the nonmaterial.

The Story Principle metaphor came, as I explained in the first chapter, from my experience as a writer. Before one of my novels comes into existence there is my imagination, which will create the story. Once it's created and published there exists the material aspect of paper, ink and cardboard (or digital file) that allows another consciousness to experience it by reading—that is, taking it in through the senses and experiencing the plot, characters and setting. So—first mind, then mind stuff shaped into story, turned into material book, and re-created as mind stuff and experienced by another mind.

Story Principle seems to me a satisfying metaphor for explaining the beginning of the Universe. (The Big Bang is too distant, cold and lifeless for my taste—and leads inevitably to the metaphor of the universe as mechanism that poses the difficult question of how mechanism managed to spawn the abundance and diversity of life that we see all around us.)

Here is the Story Principle version of cosmology: Before the Universe that feels so real to us, there was Mind, Consciousness, Imagination. (I'm using capital letters as a reminder that this is First Mind, First Cause.) All that we think of as Universe, while it has a material aspect that allows us to experience it, is—in its essence—mind stuff that evolved out of or was created by Mind, which has neither beginning nor end. The material aspect of life, the universe and everything is what we take in through our senses and experience only in consciousness. *All* is "mind stuff" that paradoxically is expressed in material form to be shared by individual consciousness.

There are other words than Mind, Consciousness or Imagination that can be used here of course—Unified Field, Void, Energy, Spirit, God, Mystery. In story metaphor it would be Cosmic Imagination, that which spins the Story of All That Is, the great Saga of "life, the universe and everything," just as my own imagination spins a novel. (And no, we don't understand the *how* of it—in either case!)

The Story Principle metaphor works on many levels. Consider the question of free will. People tend to think authors have total control over their characters. Almost any writer will tell you otherwise. Every choice you make about what a character is like limits your control of that character as the story progresses. The more real and specific you have managed to make a character, the less control you have—"I wouldn't do such a thing!" the character announces as you try to write him climbing a tree to rescue the kitten you put up there. He's right—you did already give him that fear of heights.

Most of the time character and author work together, co-creating story, and most of the time it goes pretty much in the direction the author intended. But occasionally a character rises up and takes the story in a direction the author did not intend—free will.

Think of the whole of humanity, all seven-plus billion of us, as characters in a massive, complex Saga being spun by Cosmic Imagination. The universe is, of course, the overall setting of this Saga, and our blue-green planet is the immediate background for our part of it. In the Saga, every single character is the hero, the protagonist, the central character of his or her own storyline. It doesn't matter whether that character is a terrorist, a president, a corporate executive, a newborn infant, a soccer mom, or a demented person mumbling to herself on a street corner. Each is the hero of a storyline. And the storylines are intricately interwoven.

There are no minor characters in the Saga, though every individual will serve the role of major or minor character in other people's storylines. Each is really the center of his or her own universe.

Just as each character in my novel is both created by and made up entirely of my consciousness, each of us is made up of the consciousness of Cosmic Imagination. This is the Unity part. Being *made of consciousness*, each of us has an individualized version of that same creative power. We each have our own personal imagination with which we participate in the creation of our own storyline and affect the storylines of other characters and the larger Saga in which our storylines are embedded. We may create our story in alignment with

Cosmic Imagination (going with the flow, as it were), or we may go off on a tangent of our own devising. Once our individual consciousness enters the Saga, we are at choice.

So, if we are "at choice" as you say, doesn't that make us author instead of character?

Not instead of—in addition to. We play the role of both character *and* author in our individual storyline. Here again we encounter paradox. Remember the tenet of Unity—All That Is is made up of one "substance," which we are calling consciousness or imagination. What we know of as our *self* within the storyline is only the material aspect of our larger being, not all of who we are.

Tom Toles, the political cartoonist for the *Washington Post*[2] often draws himself into his cartoons, representing himself as a tiny figure, usually at the bottom right of the cartoon, sitting at his drawing board and commenting on what he has drawn. This is one way of reminding us of the author (cartoonist in this case) self. That self is always there, whether we are aware of it or not. But it isn't really separate from the drawn figures or the whole cartoon (think character and story); the little bottom-right drawing is merely a reminder of its presence. It is another aspect of the self, functioning in the immaterial realm of consciousness.

There's yet a third aspect, also inseparable—which is Cosmic Imagination, the "stuff" of which both author and character are made, called in some spiritual traditions the God within. The three aspects of self—character, author and Cosmic Imagination are paradoxically different expressions or aspects of one thing and they can function either in unison or not. The author self, existing outside the level of story, works in alignment with Cosmic Imagination; the character self, because of the emotional impact and the intensity of the story, may or may not.

Our author consciousness is constantly creating our bit of the overall Saga in accordance with our character self's choices. *Here's the tricky part:* whether we as character choose our story consciously or not, we are always participating in the creation of it. Most of us don't realize the authority we have in our own experience. We are so immersed in the story that we don't recognize it *as* story, much less our part in creating it.

Have you ever read a book and felt horrified by the direction the story was taking, what you could see coming, frustrated that you couldn't somehow intervene and save the character? That's how most of us feel in our own stories—as if we are stuck with what's happening *to* us in a plot that is entirely outside of our control.

As we pay attention to what is happening in our story, focusing only on that experience, that perceived *reality*, it is precisely our attention and our expectation that it will play out according to the direction we see it tending that keep it going in that direction, fueled by our feelings. Most of us bump along inside our stories, creating by default, believing ourselves as unable to affect the plot as the reader of a novel.

Story Principle says that our lives in the present moment are simply the material outcome of the story we (as character and author) have been telling thus far, and that our lives in the future will be the material outcome of the story we are telling now. Few of us are aware that we can change the story we are telling in any given moment, thereby changing our future experience. But we can. What we, as character and author, have created, we can change.

Hold on. Are you saying our lives are* only *story? Figments of our imagination?

Yes and no. What I'm saying is that *everything* is story. I mentioned this in Story Principle 101 and pointed out that you could change your story without accepting that idea. But to go deeper into the uses of Story Principle in our lives it is useful to understand the concepts that underlie its working.

There are two levels to story—the level of the story itself and the level of its creation. Think of Cinderella again. To us she's a fictional character, and so only a figment of our imagination. But to herself she is entirely real! Her story is *only* story to us, but to her it is her life: her rags, her cinders and later her marriage to the prince—all real.

So it is with us. We, as characters, are embedded in our experiences at the story level, so they are entirely real to us. But we, as author, inhabit the level of creation where our experiences are only story. We exist at both levels, and it is our author aspect that makes it possible for us to change our story.

[1]Adams, Douglas. *A Hitchhiker's Guide to the Galaxy.* New York : Harmony Books, [2004], c1979.
[2]http://www.washingtonpost.com/wp-dyn/content/opinions/tomtoles/

Putting it to Work

Imagining Who You Really Are

As long as you are feeling yourself to be nothing more than a character in your story, created by some being or force over which you have no power, you will have the sensation of being caught in your story. You are likely to feel as if things happen to you that you can't help, many of which you don't like, and that much of the way your life unfolds has nothing to do with you.

Accept the possibility that you are much *more* than that character self you know by name and see in the mirror, that an important part of you exists outside the story that's unfolding around you. Because you may never have thought of yourself in this way before, you may find it helpful to create an image or description of the "rest" of yourself that will help you accept this larger identity. The task here is to imagine your whole self, the "three tiered" being (Cosmic Imagination, author and character) made up of the consciousness, light and energy of the universe. You want to be able to see yourself as a being who has both chosen and been chosen to enter this story of your life, a being with the power of the universe at your command as your story unfolds.

Whatever your physical self may seem to be, whatever image you see in the mirror or how you appear to others, those aspects of yourself that exist beyond the story level are ageless, timeless and formless. You get to imagine them in any way you choose. You can create with color, with light, with symbols. What you are after here is imagery that will take you well beyond the limitations of your day to day identity and your normal frame of reference. Your image may be as concrete or as ethereal as you like. You may borrow from traditional spiritual images, mythology, the world of animals and birds. The point is to create an image or a description that fills you with a sense of power, light, energy and awareness.

This exercise can be done in writing or in pictures, or it can be done solely in your imagination, but it is most helpful if you can distill it to a few words or images that you can call up at will. Once they've become familiar to you, you can picture the image or repeat the words in your mind every morning before you get out of bed and every night before you go to sleep, as well as whenever you are feeling frustrated, down or discouraged.

As a word person, I created an affirmation about my larger self that I began to use in the morning back when I first began exploring metaphysical ideas. I suspect just repeating the words every day has helped bring those ideas into my life. These are the words I used, based on what I felt I most needed at the time: "I am a radiant light being. I am energy; I am spirit; I am light; I am love. I have access to the infinite resources of the universe: time, money, guidance, knowledge, creativity, wisdom, peace and joy."

Readers of *Surviving the Applewhites* will recognize the first words of that affirmation as Lucille's description of Jake, the juvenile delinquent. As flakey as the Lucille character is in that book, her certainty that a radiant light being can be found in everyone, whatever their outward appearance, provides the first seed of Jake's discovery of himself. My own affirmation felt more than a little foolish when I thought those words, but has grown to be a source of real strength in my daily life.

A friend who is more of a picture person developed for herself an image of a mythical creature, part lion, part bird, part fire. She visualizes it at the center of her being—radiating its light into the darkness of an infinite internal space. Reversing the idea of an umbilical cord that extends outward, she images a cord of light that extends inward from her navel to the mythical figure she sees as her source of power. Before she sits up in bed and faces her day, she has always spent a few minutes seeing and being that fiery creature.

Whatever image or language you use as you imagine the "rest" of who you are, the idea is to stretch your sense of yourself well beyond where it has gone before. Be extravagant, remembering that you really, truly are—on the authority of contemporary science—"star stuff."

4
Mutual Reality

The real voyage of discovery consists not in seeking new landscapes, but in having new eyes.

–Marcel Proust

You say that we are embedded in a Saga with other characters who are the heroes of their own storylines And that our storylines are interwoven. Surely our inter-Actions with each other create a mutual reality over which we as individuals have no control.

There is really no such thing as a singular, mutual reality, no matter how it may seem otherwise. No matter how closely intertwined our storyline is with others in the Saga, each story—hence each reality—is an individual one, seen and interpreted from an individual perspective. Let's look at a story involving three people who share what might seem from outside to be a mutual reality.

Joe, a young father who must travel for his work, is on his way home from a business trip one day, feeling a bit guilty for being away so often from his children, six year old Ben and five year old Sarah. He hurries to one of the airport shops to buy them presents before his plane begins boarding. There he finds some oil pastel crayons. They're easy to slip into his briefcase, they're gender neutral, and he thinks that giving both children the same present is not only easy, it's also fair. Neither of them will be able to grumble that the other one got something better. He buys two boxes.

Both of the children are excited when he pulls the bag out of his briefcase. But when he gives them their presents, Sarah is disappointed. She would have much preferred a book. Ben tries out the crayons immediately. They are so much brighter and more colorful than those he has been used to, so much fun to work with, that they spur him to draw—something he has never cared much about before. He keeps drawing. Over time he becomes passionate about it and very good at it.

Forty years later, dying of cancer, Joe is gratified that both of his children are there at his bedside. He feels he has been a reasonably good father, doing his best to be there for his kids. His only regret is that Sarah has never been as easy to love as Ben. Even now, she seems prickly and standoffish, while Ben hovers close, doing what he can to make Joe comfortable.

At their father's funeral Ben, now a successful commercial artist, gives the eulogy. He tells the story of that long-ago day and the gift of oil pastels. "Dad knew me so well, better even than I knew myself, that he gave me the perfect gift that day, a gift that set me on the road to a fulfilling and successful career." Ben doesn't even remember what their father gave Sarah.

Sarah, on the other hand, remembers the day as one of many times over the years that their father favored Ben, giving him what he wanted and not knowing or caring what Sarah might want. "Dad never knew—or valued—*me*!" she whispers to the person sitting next to her. "Everything was always Ben, Ben, Ben."

We have here an apparently singular "reality" and three *stories* fraught with emotion. Each character's story on that first day was different and led to a different life experience. Having set a story in place, though they didn't know it was only story, Ben and Sarah continued to interpret whatever their father did over the years in the light of that story, ignoring or perhaps not noticing evidence to the contrary. And so their stories grew, cemented into place by each interpretation and by the interactions that provided evidence. Whoever Joe was to himself, he was a different father to each of his children.

If Ben had understood Story Principle, he probably wouldn't have felt the need to change the story he had begun to tell in early childhood. Its effect in his life was positive. Sarah, however, could have given herself a much more comfortable relationship with her father had she changed her story. Had she changed her story, her father's story, too, would have changed. A different relationship would have unfolded.

Negative feelings act as warning signs ("Bridge Out Ahead") letting us know that it would be more comfortable for us if the story we are telling changed direction. Negative feelings allow us to look at what we are telling ourselves to see if we can change it in a way that would feel better so that the experience that unfolds from that story will feel better.

We don't have to relive any past story in order to change it. In fact, reliving it keeps it operating in our experience. We don't even have to figure out why we told it in the first place. We can change a story in any moment, "selecting and deleting" any part we choose. All we have to do is locate the particular bit of story that's giving us trouble and be willing to let it go.

When Sarah first felt bad about getting crayons instead of a book, she couldn't change what her father had chosen to give her, but she could have told herself that at least her father cared enough to bring her a present. She could have thanked him for it and then, before his next business trip, let him know how much she'd like to have a new book.

That's a lot of responsibility for a five year old!

True. But stories can change at any time. Anytime during the intervening years, Sarah could have switched her attention from the slights she felt to whatever her father did that showed his caring. Having focused on one example of caring she could have begun to expect others. In other words, she could have put her imagination to work in her favor. Remember that it was her imagination that created the original idea that her father preferred Ben to her. Unfortunately, she thought it was *reality,* and that is what it became.

When we tell ourselves a story that feels better, even a little better, we change the direction of the reality that will grow out of that story. Better feeling story becomes better feeling reality. And as the reality feels better it becomes easier to tell ourselves more positive stories. While each of us perceives differently, our individual storylines are so intertwined that as our stories change, so, inevitably, do those of the people around us as they perceive our change and adjust to it.

But what of the realities we have already lived? How can we change our feelings and expectations when we have the memories, the evidence, of what has gone before?

As I've said, we can change a story at any time. The truly astounding thing is that a new story does not just change our future reality—it can actually change the past.

Let me give a personal example: My father was the kind of person who believed in organization—a place for everything and everything in its place. He even *labeled* the place! He pursued his two primary hobbies, throwing pots and making silver jewelry, in his basement, where order always reigned.

My office, on the other hand, is a minefield of piles of paper; my desk overflows with books, CD's, papers, pens, scribbled notes, candles and oddments of all kinds. As I was growing up my general messiness and my tendency to lose things drove my father nuts. He yelled. He punished. He *demanded* change. When I didn't meet his demands, he let me know in no uncertain terms that such profligate disorder was a sign of poor character. I believed what he said about poor character and beat myself up for it. But as hard as I tried to bring order to my life, I genuinely couldn't do what was so natural to him.

When my father was in the hospital, dying of leukemia, my son arrived home from college for summer vacation and went to visit him. Dad and RJ had always gotten along, Dad being especially taken with RJ's combination of creativity and intelligence. The two had a good visit. When RJ left, Dad said, "That's one great kid you've got there." I nodded my agreement, but added that in the less-than-24-hours RJ had been home, the mess in his room had gotten so bad that it was almost impossible to open the door.

Dad smiled. "The Great Ones are *all* like that!" he said.

It took a moment for me to take that in. The Great Ones are *all* like that. My first reaction was rage. How could he so casually admire my son for the very trait he had shamed me for my whole life? Why could he never have said such a thing to me? What a difference that would have made! It was all I could do not to storm out of the hospital.

At home that night I continued to fume about the injustice of it all, until—in the middle of the night—it dawned on me that what Dad had said about RJ retroactively applied to me. The Great Ones are *all* like that! The pride he had never shown me, now directed to my son, suddenly became available to the child I had been. Here is the astonishing thing: with that realization, *my childhood changed.*

After all, there was nothing left of that childhood and the frightening, overbearing, demanding father I was never able to please, except memory—*story.* Changing the tyrant in that story to a man who could see and admire the potential in a bright, messy, creative kid changed everything. The father and daughter who faced Dad's death in the week that followed had become different people from the father and daughter who had raged at each other for decades.

Story, past or present, can change at any moment!

Putting it to Work

This exercise is designed to remind you that your story is not the only story going on in any situation involving another person. Remember that none of the exercises provided here are tests of anything. You don't have to worry about doing them wrong. They're just ways to play with and become used to the ideas of Story Principle.

Exploring a "Mutual Experience"

> Choose a problematic relationship in your life (here again, you might want to begin with one that doesn't carry too big an emotional charge). Let's say, for instance, that there is a particularly grumpy, hostile, humorless clerk at the post office you go to most often. Write a brief description of the other person, including things she has done or said that have made your interactions unpleasant. Have some fun with it—writing dialogue that you either remember or make up to fit the image of the character. Think, for instance, of the "Soup Nazi" on the Jerry Seinfeld television show.
>
> Then, try doing the same exercise from the other person's point of view. Imagine the same incident and the story that person would tell about you. Again, have some fun with it. Think of this task the way a novelist would—you're

trying to get inside the other person's mind and to see yourself only as the other person might have seen you. It doesn't matter whether you're "right" about this person's viewpoint, it only matters that you imagine it as clearly and as believably as you can.

Next try to remember whether you've ever had an experience with this person that doesn't fit the negative story you've told. If you can't come up with any, see whether you are able to reinterpret something from the experience you've already recounted in a different, more positive light. Or imagine a reason for this person's attitude. What is her home life like? What might have happened to her to give her such a sour attitude?

The point of this exercise is not to change your story about the other person (though that could be a possible outcome) but just to consider how an apparently mutual reality may be viewed very differently by its participants. It's a way to help you remember that each of us is both a central character in our own storyline and a minor character in someone else's.

Remember you are just an extra in
everyone else's play.

–Franklin Delano Roosevelt

Remembering this can go a long way to changing both the small stories you tell yourself as you move through your day interacting with a variety of people whose interests and intentions may be quite different from your own, and bigger ones that you may never have considered from anyone else's point of view.

When you get used to doing this from time to time, it may help you begin to revise the stories you are telling yourself about the people in your life. As you change those stories the people will necessarily adapt to your new attitude and behavior. Once we are able to grant others the right to have a different story than our own, relationships begin to change.

If the postal clerk doesn't get any nicer, at least you're likely to be less irritated by her attitude in the future. Sometimes as we change our story about a maddening relationship, that relationship vanishes from our experience entirely. Perhaps she'll get transferred to an office on the other side of town!

5
Belief

I could not have predicted that I would
write an entire book
about the fact that beliefs have physical
repercussions...

–Herbert Benson, M.D.

A little more than ten years ago the members of the writer's group I belonged to in Ohio were complaining, as we often did, about the difficulty of making a living as a children's writer. Only two of us were trying to maintain ourselves solely through our writing, and while we didn't have full time jobs like the others, both of us had to supplement our income with other activities. What we really wanted was to be able to live by writing.

I'd been reading some metaphysical literature that suggested using affirmations to change your experience, so I announced to the group that I was going to try it. "I am a successful children's writer!"

I said, pledging to repeat it as often as I could remember. A lot of people would have said I already was successful, since I had been publishing steadily for twenty years, but my writer friends and I had a different definition of success.

I remember driving my old un-air conditioned Dodge Colt down the street on the way home from our meeting, repeating my affirmation over and over. After a while I started to laugh. "If you're so successful," I asked myself, "how come you're driving this car?" I nearly stopped the experiment then and there. The affirmation felt like a lie. But I decided to give it a little more time. Over the next weeks and months, I would occasionally remember to repeat it, sometimes as often as ten times in a row. Usually by the time I finished, I was giggling. Saying it over and over wasn't helping me to believe it. But the metaphysical books were telling me that in order to create my own reality, I needed to believe in what I was saying.

The Power of Belief

You're no doubt familiar with the *placebo* effect in medicine—the example we think of is a sick person being given a sugar pill that contains no medicine at all, and being cured by it. Modern medicine has to take the placebo effect into account whenever it tests a new drug, but for the most part the phenomenon is set aside as something of an irritating anomaly. It doesn't fit easily into the old paradigm of scientific materialism, giving evidence, as it does, that nonmaterial mind is able to change the material reality of the body.

Herbert Benson, M.D. author of *Timeless Healing, The Power and Biology of Belief*[1] has studied the placebo effect for several decades, and has noted that the human body has a "propensity to turn a person's beliefs into a physical instruction." (p.21) In the cases he has reviewed 70-90% of documented therapeutic effects can be accounted for by the placebo effect (which he prefers to call "remembered wellness") rather than the medications or procedures provided. Conventional medicine tends to attribute only 20-30% of successful outcomes to the placebo effect. But more importantly, conventional medicine tends to ignore the fact—shown by the very existence of the placebo effect—that our beliefs affect our physical experience. Benson is among the minority of medical researchers seriously investigating the power of belief and considering how to use that power to improve our health care system.

While it may be comforting to know that believing in the medicines we take can assure that we are more likely to benefit from them, the effects of belief are not always positive. Patients unknowingly receiving a sugar pill may exhibit the negative side effects of the medicine they believe they're taking. This is an example of the *nocebo* effect, where a belief in harm brings it about. The *nocebo* effect is what kills believers in Voodoo who have been cursed by a witch doctor. Benson quotes Dr. Herbert Basedow's 1925 eye witness description of such a death caused by an Australian aborigine witch doctor's "pointing a bone" at a member of his tribe: "The man who discovers he is being boned, is indeed, a pitiable sight...He sways backwards and falls to the ground and after a short time appears to be in a swoon; but soon after he writhes as if in mortal agony and covering his face with his hands begins to moan...His death is only a matter of comparatively short time." (p. 40) We tend to think such a death is an example of woeful ignorance—but the nocebo effect is also found in our own world.

Bruce J. Lipton, in *The Biology of Belief*,[2] gives the example of a man who was diagnosed with esophageal cancer at a time when that diagnosis was considered a certain death sentence. He was hospitalized and treated by medical personnel who "knew" that he was dying. He died, as expected. But when an autopsy was performed it was discovered that he had only a couple of spots of cancer in his body, and no trace in his esophagus. His doctor is quoted as saying "He died with cancer but not from cancer." (p. 143)

Like the Voodoo believer who dies of a curse, the aborigine who dies when a bone is pointed by a witch doctor, the modern medical patient may die of a diagnosis. The more authority the patient grants the doctor, the more friends and loved ones accept the diagnosis and behave as though the patient is dying, the more sick the patient feels, the more likely he is to fulfill the prophecy of that mistaken diagnosis and die. Belief can literally wield the power of life and death.

A major factor in creating the beliefs that make the placebo and nocebo effects possible is *authority.* The doctor, the institution of modern medicine which backs him or her up, and the science from which medicine emerges are all given the status of authority in our culture. What comes down to us from these authorities, particularly when it is so obvious that they themselves believe it completely, we are likely to accept.

Anyone who seems to know more than we do or anyone who has power over us assumes the status of authority. Authority may also be conferred by numbers—what is believed by most of the people around us is hard to resist. So, although belief is usually created through a combination of factors, of which authority is only one, its power can be considerable.

Experience is another belief-inducing factor. Our own current and previous experiences build belief, as do experiences we observe in others. A powerful enough experience—like burning our fingers in a flame—convinces us, essentially permanently, that fire is hot and must not be touched. In fact, we categorize that not as a belief, but as fact—truth. Experiences fraught with emotion, whether positive or negative, tend to build belief more quickly and easily than neutral ones, and our own experiences are usually more compelling than those of other people.

It isn't necessary, by the way, for an idea to be *true* to be believed. Hitler's propaganda principle was that if someone in authority tells a Big Lie often enough, people will come to believe it. Authority and repetition work powerfully together.

Repetition on its own, without the support of authority and experience is not extremely powerful. This is why affirmations alone aren't likely to quickly create belief. If we don't have a sense of our own authority and if we don't have any experience of what we are affirming, affirmations tend to set up an internal argument that can actually work against what we are attempting to affirm. Arguing with ourselves does not give us the focused consciousness that makes belief powerful enough to repel anything that contradicts it.

Back when I was repeating the affirmation that I was a successful children's writer there was no established authority telling me that I could make a living as a writer. In fact, the only authority I had ever seen (a survey taken of children's writers in the 70's) had pointed out that only 3% of children's writers were able to do that. And while I knew some who were doing quite well, I knew a great many more, particularly those who wrote the kinds of books I did, who were struggling as I was. My own experience, after twenty years of publishing, was not giving me support for believing in what I wanted. And since I had not yet discovered Story Principle, it didn't occur to me that my experience was the concrete result of the story that I (and the other writers I knew) had been telling all along.

Nevertheless, at the beginning of my experiment affirming my success as a writer, even while I continued to argue against the idea, I did begin to notice a small effect. The first time I called my agent after affirming that I was a successful writer, it was easier to insist on being put through to her. Repetition may not be the *most* powerful factor in building belief, but it's a useful starting point.

For years I occasionally repeated my "successful children's writer" affirmation, never quite getting over the tendency to argue the point. But at the same time I went on reading metaphysical texts and new science that began to provide some authority for the idea that creating one's own reality was possible. Still, experience was not there to support me. The metaphysical books I read went on touting belief, but I hadn't yet found a way to concoct it for myself.

Now here's the thing about stories—*they don't require that we believe.* All a story requires of us is a "willing suspension of disbelief." There is a huge difference! It takes time and effort—authority, experience, repetition—to fully establish belief. It takes only *willingness* to suspend disbelief.

And here's the thing about the Universe—it's on our side. (Trust me on this for the moment. We'll get to it more fully later.) The Universe is so thoroughly on our side that it will manage to squeeze the fulfillment of our story through the crack we've made by suspending disbelief, even if we are nowhere nearly ready to actually *believe* in the story we have told.

[1]Benson, Herbert, M.D. *Timeless Healing, The Power and Biology of Belief.* New York: Scribner, 1996.
[2]Lipton, Bruce, Ph.D. *The Biology of Belief, unleashing the power of consciousness, matter & miracles.* Santa Rosa, CA: Elite Books, p. 142.

Putting it to Work

Creating a Hero Self you can believe in

In the Old Testament of the King James Bible Moses asks God's name and God answers "I am that I am." The power of this expression of ultimate beingness suggests that whenever we begin a sentence with "I am…" we are making a statement of identity which carries that same power. We create ourselves by the way we describe ourselves, the way we finish the sentence.

This exercise is designed to focus your attention on your ability to create yourself as the central character, the "hero" of your own story in whatever way you choose, with whatever traits, attributes and gifts you can allow yourself to believe in. No sense trying to give yourself super powers if you are clear that super powers exist only in comic books.

Good stuff

Divide a sheet of notebook paper into two columns. Title one of them *Good Stuff I've Done* and the other *Good Stuff I Am.* Start making a list under whichever heading seems easier. Note that both lists are positive. If negative things come up, let them go. They don't count here.

You may be able to quickly make two nice long lists. Or you might encounter something of a struggle. If it's easier to list good things you've done than good things you *are* you will be able to derive from the list of what you've done the attributes that belong in the other column. If you've donated to good causes you might say "I am generous." If you've overcome serious illness, you might say "I am brave," "I am strong," "I am healthy." If you've come up with a new way to do something you could say "I am creative," "I am smart." Only you can decide which attributes fit which actions.

If you are having difficulty you might try looking back over your life as if you were a stranger engaged in a treasure hunt, a stranger who will win a prize if his lists are longer than anyone else's. Did you stand up to a bully to help a friend when you were in the second grade? "I am brave." *Any and all evidence counts!*

Not so good stuff

When you've come up with as many additions to your lists as you can think of (it may be worthwhile doing this exercise over several days so you have plenty of time to process and remember), get out another sheet of paper and write the heading, "Useful or Interesting Flaws." Heroes are multi-dimensional folk—it wouldn't be much fun to read a story, whose central character has only good traits and does only good stuff.

For example, my sixth grade teacher, the dreaded Miss Shreve, told my mother I would never amount to anything because I was interested in "too many things." There aren't many traits I now value more! And while it has made for some difficult times, I eventually realized that being "too emotional" and "too sensitive" were extremely useful traits for a writer. As short story writer Shirley Hazzard said about writing, "The state that you need to write is the state that others are paying large sums to get rid of."[1]

Being stubborn allowed me to stand up against my father's version of who I was or should become. As for my messiness, forgetfulness and disorganization—well, if those aren't overwhelmingly useful, at least they can be interesting.

You may now wish to use your three lists and write a character sketch about yourself as the hero of your own story. The idea is to be able to feel about yourself the way a writer feels about the central character she has chosen to write about. If there are to be many books or stories about this character (think Agatha Christie and Hercule Poirot, or J.K. Rowling and Harry Potter), she wants to really, really like him. They're going to spend a lot of time together!

You may find yourself coming back to these exercises from time to time as you go on working with Story Principle. Your story about who you are is almost certain to change.

[1]Keyes, Ralph. *The Courage to Write, How Writers Transcend Fear.* New York: Henry Holt, p. 84.

6
Suspending Disbelief

I am too much of a skeptic to deny the possibility of anything.
–T.H. Huxley

In 1998 the Ohio Arts Council awarded me a summer writing residency at the Fine Arts Work Center in Provincetown, Massachusetts. I was provided with an apartment and a small stipend. From the first of June until the last of August, I had no responsibility to do anything except write. It felt like a miracle—something I had been dreaming about my whole writing life. So I dubbed it my Miracle Summer Residency.

All summer, whenever anything seemed to be going wrong, I told myself, "This will work out perfectly and quickly—because this is my Miracle Summer Residency." And all summer that turned out to be true. Whatever seemed about to present a problem would get solved almost instantly.

Synchronicities (that felt like mini-miracles) abounded, most of them minor, but some of them distinctly memorable. One day, for instance, I decided to take an uncharacteristic break from the writing and ride my bike into town for a mocha frappacino. As usual the sidewalks of Provincetown were packed with people moving from shop to shop, gallery to gallery. I noticed what looked like a familiar figure heading into a shop. It was Karen Hesse, a children's writer from Vermont who had won the Newbery that year (we'd had dinner together at a conference the week after she learned of the award). She had come to Provincetown for the afternoon with her husband and children. We chatted for a while, catching up on each other's lives, chuckling with amazement at our chance encounter. A minute either way and it wouldn't have happened.

Midway through the summer the Today Show contacted me to see whether my son and I could be on the show a few days after my residency ended, to talk about highly gifted children. I agreed, but then fretted that I had nothing to wear—I had brought nothing for my summer on the beach that would be appropriate for a Today Show appearance. A friend was visiting who knew something about dressing for television and agreed with me that nothing I had would do. She helped me find and buy a perfect blouse and when I complained that I couldn't afford the matching pants she assured me that

I'd be sitting down. "The pants you wear won't matter." And so it was.

All in all, though I missed my husband, it was a glorious summer that fit the label I had given it perfectly. I finished my novel, *Ordinary Miracles,* and wrote about half the essays for a nonfiction book for adults. I rode my bike, walked the beaches, collected beach stones, read more metaphysical books, watched sunsets and sunrises over the water (Provincetown's geography provides both!) and basked in my summer miracle.

When I got home that fall, my life returned to normal. When things started to go wrong, they went wrong. Money, as usual, was tight. We had decided to move to be near my husband's family in North Carolina, and I wasn't at all sure we could handle the cost. He was semi-retired, and I was a writer—a children's writer. Could we even get a mortgage?

Setting Intentions

But I'd been reading those metaphysical books. And I'd had that Miracle Summer Residency. Maybe, I thought, there really was something to having positive intentions, positive expectations. We found a wonderful house on a little lake in a big woods only fifteen or twenty minutes from downtown Charlotte.

Unfortunately, the house was too expensive. Vastly too expensive! My husband, raised in Christian Science, has been a lifelong optimist. He was all for focusing positive intentions. We decided to *intend* getting the house. The owner brought the price down so far that we could handle it, and the bank cheerfully gave us a mortgage, but the closing date was only a month away and we hadn't begun to deal with selling our other house. I decided to *intend* to sell our old house in a week (my husband thought that a tad *overly* optimistic, but I suspended my disbelief)—and we sold it in a week, to someone who could pay cash so that everything would work out for the April 15 closing on the new one.

April 15, 1999. Tax day. Every other year that date had loomed on the calendar, stirring me to near panic by mid March. Would we be able to afford to pay our taxes?

Writers have no way of knowing how much income they will have in any given year—it all depends on how many books have sold. Worse, there's no automatic withholding from a royalty check. The question was always whether I had managed to save enough to cover the tax bill. That year, even with the cost of the move, we managed to scrape by. But I was tired of the stress.

A New Story

The day after we moved into our new house, looking out the floor-to-ceiling windows into the trees and down to the lake, I had a spectacular idea. If I could win a Miracle Summer Residency from the Ohio Arts Council, why couldn't I win a Miracle *Year* Residency from the Universe? I imagined the details, based on my Provincetown summer. This residency would allow me to write in the wonderful house we'd just bought overlooking the wonderful lake! I would have my husband with me so it wouldn't be so lonely. I could have my dog and my cat and I could grow my own tomatoes! More importantly, the "Miracle" in its title would assure me that during the year of the residency anything that seemed to go wrong would work out quickly and easily, just as things had during the Provincetown summer.

As the final perk, I added a stipend to the story equal to whatever we needed during the year. When my old rational-materialist voice questioned the stipend I assured it that if the Ohio Arts Council could be counted on to pay me for a summer, surely the Universe, with its vastly greater abundance and its track record of endurance, could be counted on to cover whatever costs might come up in a year. I didn't even need to labor over an application—I just had to tell the story that the Universe had granted me the residency I had designed and suspend my disbelief. It worked. I wrote steadily, reveling as much in our natural surroundings as I had in the beaches of Provincetown, and money came in when it was needed, often in surprising ways. The next year, on April 15, I renewed my residency.

Freedom

For three more years I renewed the Miracle Year Residency each April, and each year the Universe, even when it seemed most unlikely, came through—sometimes at the last minute. I should say here, in the interest of full disclosure, that this newly minted story didn't give me three years of stress-free living! I had a bit of trouble remembering it. But every so often something would remind me, and I would retell it firmly. The Universe wasn't having trouble providing from its abundance—*I was having trouble remembering the story and allowing it to unfold.* But each year that I told the story we did, in fact, have what we needed.

Then, in January of 2003, I won the Newbery Honor for my novel, *Surviving the Applewhites,* and my success as a children's writer became undeniable, even to me. There was no longer a question about making my living as a writer. One of the members of my Ohio writer's group remembered my affirmation experiment and called me. "It worked!" she said. "Next time figure out a way to make it work faster!" But I knew it hadn't been the affirmations alone that had worked—I had actively begun telling new stories and suspending my disbelief.

Having won the Newbery Honor and finding that money was no longer such a source of fear, I made yet another story—one I might not have had the nerve to create before. Suspending disbelief requires only willingness, but willingness comes more easily when the story has consistency, when it holds together and has some connections with what has come before. Based on the success of the Miracle Summer and Miracle Year Residencies, I invented the Lifetime Freedom Foundation—a foundation sponsored by the Universe, whose mission it is to free humanity from its self-imposed chains and limitations. Then I invented their Lifetime Freedom Fellowship and gave it to myself. I documented this new storyline by writing myself a letter from the Board of Governors. The Fellowship guaranteed that I would have the freedom to do what I needed to do not just until the next April 15th, but for the rest of my life. I put the letter in a safe place, where I could get it out whenever I needed to be reminded of its promise.

I shared these stories with only a few carefully chosen people until they had materialized in my life in obvious ways—a lesson I learned during my Miracle Summer Residency. (The writer's group I met with on Cape Cod—intelligent, well-educated and talented women—kept wondering what sort of emotional crisis could have led me to consider the sorts of far out metaphysical ideas I was talking and writing about.) But even after my financial life had changed radically and undeniably to fit my new stories, a few of the friends I told about those stories scoffed at the idea that I had created the changes myself. People might be amused at the "coincidence" of changes appearing after someone has told the story of those changes, but the old way of looking at life as something that happens *to* us, with control relegated at best to how we cope, is so ingrained in us all that it is hard even to suspend disbelief if we haven't taken the step by step journey from the old story through our chosen revisions to the new reality ourselves.

This is why this book encourages you to try Story Principle for yourself instead of merely reading about it, no matter how many examples of its success I can provide. It is only when the stories we begin to tell for ourselves begin to show up in our life experience, changing as we change them, and appearing in surprising, even completely illogical ways, that we are likely to accept the possibility that it *really is our new stories* that are changing our lives.

Little by little we begin to accept our own authority (author-ity) and revise more bits and pieces of our story until the complete formula for creating belief is in place—authority, experience, repetition. This is the way Story Principle is transformed from radical concept to a possible way of life.

It's fun. It's playful. You tell yourself a positive story and then sit back, smile, and say, "this is a lovely fantasy I'm willing to entertain." It's true that it can sometimes take a while, but as long as you aren't mentally and emotionally contradicting it every other moment, your story becomes your life.

This sounds way too easy.

The truth is that it's more simple than easy. The processes I've just described were more problematic than the telling makes them seem. The pessimistic, rational-materialist self I used to be did not vanish—she kept raising objections. And the monsters of Lack and Scarcity kept getting out of my nightmare closet. On those occasions when the money appeared at the last possible moment I had plenty of wakeful nights. Fear is a tough adversary. I would take two steps forward and one step back—over and over. But as long as I went forward again, I made progress. And the more progress I made the more easily things went.

You don't have to maintain your story every single moment. As Ernest Holmes said, you have only to think positively about what you wish to bring into your life 51% of the time.[1] As you work with Story Principle, remember that it's a process. Buddhists call their spiritual efforts "practice."

It is not a matter of faith; it is a matter of practice.
–Thich Nhat Hanh

The more you practice telling new stories and suspending disbelief, the better you get. Setbacks are normal, even useful (you can learn from them). Tell yourself the story that it gets easier. And it will. That's how it works!

[1]Holmes, Ernest. *Creative Mind.* Burbank, CA: Science of Mind Publishing, 2005, p. 19. [Original copyright, 1918]

Putting it to Work

Always remember that you're working with a principle that doesn't distinguish between fiction and reality. Everything is story, whether it has already become real in your life or is on the way to becoming real.

In 1740, when Samuel Richardson's novel *Pamela* was published in England, novels were a new form of literature, and many people considered *Pamela* morally unacceptable, not because of anything salacious contained in its pages, but because it was a "lie." As a graduate student in English literature, I remember thinking that the concern of those early novel readers was a little like the way I felt about the "big lie" I had told in the fourth grade. *Calling a good story a lie makes it hard to enjoy.*

Just so the difficult part of Story Principle is not so much coming up with new stories. The difficult part is keeping ourselves from dismissing them as lies. It isn't the Universe we need to convince, it is our own consciousness. We need that willingness to suspend disbelief. The following are some proven ways to help achieve it.

Documenting a New Story

> The first document for your story may be the story itself. Actually writing it down helps make it concrete for you. You can just *tell* yourself the story, or jot yourself some notes about it, or you can write the whole story in a notebook or

journal, complete with the feelings that would accompany its appearance in your life.

In spite of being a writer, I don't often write out my whole story, but I do often choose written documentation, like the letter from the Board of Governors of the Freedom Foundation, to support it. Documenting a story like this is play, a kind of "let's pretend" game. In much the way I used to create the most authentic-looking pirate treasure maps I could manage as a child (using yellowish paper and burning the edges in a way that made them look old to me) I like my documentation to look as real as possible. For the Freedom Foundation letter I searched the internet for an image that would work as a logo, and found the Chinese characters for freedom. I created a letterhead for the foundation, choosing an interesting font and integrating the foundation's name and mission statement: "Dedicated to freeing humanity from its self-imposed chains and limitations," with the logo. Then I wrote the letter, printed it out and added a nearly illegible signature for the President of the Board of Governors.

I did not go so far as to mail it to myself, but it would have been fun to go to the mailbox and find an envelope from the Freedom Foundation. If you're documenting a story with a letter to yourself, it might be worth the stamp to actually send it.

There are all sorts of stories you can document in writing. You can manufacture an award certificate (complete with gold seal) that documents a success you wish to achieve, write yourself a check for something you want to sell or accomplish (Jim Cary tells the story of writing himself a check for a million dollars as payment for a role in some future movie and carrying it in his wallet until the real one came into his life). You can write a love note from your ideal lover, a letter of apology from someone who has done something that hurt you, or a job offer detailing the perfect job. You can create a deed for your dream house or an offer from a buyer for your current house. Any story that can be supported in writing allows you to generate a document for yourself.

Note the words "for yourself." You aren't counterfeiting—whatever document you choose to create is for your own eyes only. It's better not to tell anyone what you're doing, unless you have a "Story Fellowship" person or group also practicing Story Principle. You aren't likely to be helped to keep a positive focus by people telling you you're nuts to waste time on such foolishness.

When you've created your document[s], put them in a safe place where you can find them and look at them again from time to time while you wait for the story you have documented to come into your experience. Should you be assailed by doubt, get them out and look them over, allowing yourself to feel the way you'd feel if they were real. You may store your documents in an old shoe box, a fancy box titled Treasure Chest, a file folder or one of those "Important Papers" faux leather folders to remind you of how important your story is.

What You See Is What You Get

Not everyone is a word person. If images appeal to you more, you may choose to document your story with pictures instead. If you're an artist, by all means paint or draw or sculpt images that fit your story. Or get a photo album, label it your "Treasure Book" and collect images from magazines or the internet to include. Because I like both words and images, I have an album that includes both—the Freedom Foundation letter is there between a picture of a kitten looking into a mirror and seeing the image of a lion, captioned "What matters most is how you see yourself" and a photo from a nature calendar of a cave behind a waterfall that is my ideal meditation setting.

Just looking at that photo cheers and calms me. And often, when I meditate, I begin by imagining myself there, feeling the spray from the waterfall as the water shimmers with rainbows.

If an image is strong and clear enough in your mind, you don't necessarily have to get it on paper. A few years ago a tax preparation company had a commercial on television that showed paper money flying through the air from every direction (like birds flocking), to represent the huge tax refunds they could get for people. That image stuck firmly in my mind and I replayed it whenever I felt my bank account was shrinking faster than I liked. Later, when I was out walking the dogs in the woods one day, I meditated on a boulder and then began to wonder how many leaves (on trees or bushes) I could see just from where I sat. Millions, I assumed. It was a great image of abundance, so I sent the message to the universe and myself that if I wanted, I could have a dollar bill for every leaf within my vision at the moment. When fall came and I walked to the boulder through the heavy leaf cover, it was an even more vivid image of abundance, scuffing my way through all those dollar bills.

Some people suggest creating collages or "vision boards" with images that fit your story and putting them up where you see them on a daily basis. That doesn't work well for me because images that are in front of me all the time I tend to quit noticing, but I often get out my treasure book and leaf through it, feeling good about the once-just-story images that have become real for me (like the newly remodeled kitchen and the laptop computer) and the images that are yet to show up in my experience. The point is to do what works for you. And keep in mind that however serious the story you are telling might be, the telling of it and the documentation should make you feel good—the way playing pretend games made us feel good as kids.

Focus on the what and the why and not on the how.

Try not to spend any time thinking about *how* your documented story is going to find its way into your life. Your "grownup" self, your logical, rational mind, even if it doesn't scold you for telling lies, may begin immediately to provide you with reasons why your story can't possibly happen. At the very least it will remind you of *real* present circumstances that are anything but the story you've documented. Pretend, as you may have done from time to time as a child, that you don't hear that adult voice and go on playing the game. This is important! This is not the time to engage the logical mind. The more you focus on the how, the more you will either distrust the story altogether, or set in stone whatever way you are able to see it happening. Either interferes with the process.

If you adopted my leaf image for financial abundance and decided that the only logical *how* for so much money coming to you would be to win the lottery, you would be focusing your attention not on what you want, but on a single, external way of getting it. That would limit the activity of the Universe to work through the infinite complexity of the Saga. Trust that your author self, in alignment with Cosmic Imagination, has a far broader and deeper perspective than you do, as well as "methods you know not of." Let the how of your story be a surprise. That element of surprise is a big part of the fun!

To keep your mind off the how, spend some time thinking about *why* you wish to allow your story to unfold. If you like, write down all the reasons you can come up with. The more why's you find, the more you will understand about what this story means to you and the clearer the story will become in your consciousness.

A story told with corroborating detail, imagined with feeling, and intended for clear reasons, has become a reality as soon as it's told, just as a novel has become a reality the moment it's written, long before it appears on bookstore shelves. Your story exists in your consciousness and, therefore, in the consciousness of the Universe. Just as a successful author sends off a finished manuscript to her waiting editor, knowing that the process of putting the book out into the world has begun, you can then go about your daily life, releasing your story to the next part of the creative process, which is not your job.

There is usually some interaction between writer and editor as the publishing process goes forward. You may get an intuitive sense that something about your story needs revision, or you may just want to check in from time to time, revisiting your story and reminding yourself that it's getting closer and closer to appearing in material form. Meantime, you may choose to spend some time with another story thread or just pay attention to all the good things that are already present in your life even though the story hasn't shown up yet. The idea is to go on about your business as confidently and peacefully as possible. Focusing on *not* having it yet slows down the whole process, because then the story that you are sending out to become real is that you don't have what you want!

In academic writing when someone quotes something that hasn't been published yet, it is referred to as *in press.* Practice telling yourself that your story is *in press.* It's real, it's quotable (useful in the world) even though it hasn't shown itself in its final material form.

7
Multiple Story Threads

If we do not change our direction we
are likely to end
up where we are headed.
–Ancient Chinese proverb

Our personal story is not a monolithic, singular tale, but a weaving together of many contributing storylines. We have stories that we are telling in the moment and stories that are running in the background of our consciousness. The stories that run in the background help form the ones we are telling currently, often without our realizing the effect they are having.

Some background stories can distort our view of current reality and our way of imaging the future. They may influence what we allow ourselves to want, to dream, to wish for. Some of these stories are so basic, so much a part of us, that they are hard to notice, let alone change. They can feel like who we are. But every one of them can be changed. When we change a background story for the better, everything else changes more easily.

Among the storylines we live out daily are Small Stories, Big Stories and Foundational Stories. The easiest ones to find and change are the Small Stories, the sorts of stories the first Putting it To Work section focused on changing. They come and go anyway. We don't need much documentation and suspending disbelief isn't difficult. The hardest to recognize are the Foundational ones. They are the deep background stories that have been with us the longest; they feel most real, most comfortable, and safest. Even if our Foundational stories hurt us, their familiarity can make them feel safer than a venture into the Unknown.

One reason small stories are easier than big ones to change is that we aren't very deeply invested in them. There isn't a lot riding on the outcome. We can tell a small story that feels good, making it just general enough to avoid an immediate setback, and then watch it materialize in our experience.

There's also a bit of an advantage to being a beginner. It's a little like writing a first novel. We're just trying it, aware that it's a challenge and we're not intending to give up our day job just yet. The more relaxed and casual we are about the process, the more playful our attitude, the better.

The more we tell small stories and watch them unfold in our lives, the more we begin to expect the process to work. Then, as we begin to expect things to work out for us, we are likely to discover we have begun to change a slightly bigger story about the way our world works.

My husband hates to get snarled in traffic. He's the sort of person who would rather travel an extra twenty miles on an open road than creep five miles bumper-to-bumper. And while he admits that when I'm with him and telling my story about getting where we're going in perfect right timing we can usually avoid unreasonable tie ups, he is not so sure about making this Small Story business work for himself. So—like most people in Charlotte—he plans his day to avoid rush hour whenever possible.

One day recently he reminded me that I'd planned to do a quick run to the grocery before dinner, but it was now ten after five. "No sense going now," he said, "it'll take you forever." I finished the paragraph I was writing—which happened to be about small stories—hibernated my computer, and waved to him as I picked up my keys.

With small stories so fresh in my mind, I set one in place as I drove away. *Smoothly flowing traffic, good parking space, everything I need available, quick checkout, smooth trip home.* Not only did I get the flowing traffic and the good parking space, but one of my favorite smaller shopping carts (usually not available when the store is busy) was stationed right at the front of my parking spot, waiting for me. And though the store was crowded and there were long lines at the cash registers, somehow as I came to check out, the closest cashier had only one customer, who finished his transaction just as I came up behind him. I was out of the store in a couple of minutes.

"Show off!" my husband said to me as I put the grocery bags down in the kitchen a little later.

"Just practicing," I told him. For me, as for everybody, small stories are easier than the others! We can come to be masters at creating positive small stories, while we continue to live with negative big ones. But changing small stories gives us evidence that story *can* be changed. Little by little our confidence grows and we may reach the point of looking into the big ones. If it works for small ones, maybe these can be changed, too.

Big Stories

Big Stories involve more time, more people, more of what we care about. They're harder to recognize as story, being based on more experience and more habit of thought. We're emotionally involved in them and they may be very important to us. Most of our big stories aren't ours alone. Other people are telling them, too, including some of the authorities in our lives. Because so many people support them, these big stories have built up an impressive amount of evidence. As they have been told and retold, they have become both past and present reality.

Here is a Big Story example from my own life. My maternal family line has early-onset Alzheimer's, which we are told is one of those 50-50 genetic diseases. If you have a mother or a father who had it, you have a 50-50 chance of having the gene that causes it. (This is not the case with the other form of Alzheimer's that comes in old age and which all people are said to have roughly equal chances of getting.)

In my mother's family the disease worked out to be exactly 50-50 for two generations in a row. There were four children in her mother's generation; of the four, two were stricken with Alzheimer's. Alzheimer's was not known then. It was called senility and no one understood the cause. My grandmother spent her last years in a mental institution—a particularly ugly story. Half of my mother's generation (six children who made it to adulthood) also got Alzheimer's—my Aunt Kathy, my mother, and my Uncle Lloyd. Between myself, my siblings and my cousins, there are eleven of us in the next generation being told that 50-50 story.

This is a Big Story for several reasons. One is that the medical community (*Authority!*) supports the genetic piece of the story, with the expectation that each child has a 50-50 chance of having the gene and also that whoever has a gene for the disease is pretty nearly certain to get it. Technically, they have to say "nearly certain" because there are cases in which one identical twin gets early Alzheimer's and one doesn't. Part of the current research is to find the "trigger" that turns the gene on in the hope that people who know they have the gene can avoid the trigger and so escape the disease.

The trigger is assumed to be a material one—like the aluminum in deodorants or cooking pans that people got worried about a decade ago. In spite of what is known about placebo and nocebo effects, *belief* isn't on the list of possible triggers. Deciding not to get the disease isn't generally considered a useful strategy.

Alzheimer's is also a Big Story, of course, because of the emotional charge it carries. For a family that has lived with this disease it is hard to imagine a worse experience. I won't tell that story here, but I wrote part of it for the charter issue of *More* magazine[1] published under the title, "The Dying of the Light." As difficult and painful as our father's death from cancer was for all of us, Mom's twenty-two years of Alzheimer's was infinitely worse.

Biology of Belief offers Lipton's contention, supported by evidence from cell biology, that the biological community is not correct in its focus on genes, on DNA, as the primary determinant of human physiological expression. On page 84 he says, "The notion that only physical molecules can impact cell physiology is outmoded.

Biological behavior can be controlled by invisible forces, including thought, as well as it can be controlled by physical molecules like penicillin, a fact that provides the scientific underpinning for pharmaceutical-free energy medicine."

Is he right? In the context of Story Principle (*everything is story*) that question is very nearly meaningless. He has a story that differs from the story of most of his medical colleagues. I choose Lipton as my authority rather than the rest of the medical profession because I far prefer his story to theirs! To change a Big Story one needs all the help one can get.

Quite a number of years ago when I mentioned my fear of getting Alzheimer's (my sister, brother and I were probably among the first people in America to own the book *Final Exit,* the Hemlock Society's book about how to commit suicide, which should tell you how strongly we felt about avoiding what happened to our mother), a friend of mine said, "Take it out of the law!"

This was an expression she had gotten from her Rabbi grandfather. Speaking something out loud put it into "the law," that aspect of the Universe which turns nonmaterial thought into material reality, bringing it into your life, whether you want it or not. If you don't want to experience something, then, don't talk about it! That's what her grandfather had told her.

I vowed then and there to change how I spoke about Alzheimer's. From then on I would only say, about the

50-50 history in the family, that I had decided not to participate. Several years later a healer I met told me that I had "outgrown my genetic pattern" and was not a candidate for Alzheimer's disease. Was she right? Again, the question about right or wrong is not particularly meaningful. It is a story far more useful in my life than the fearful one the medical profession would have me believe about my family's genetic heritage. Given the choice between a placebo story and a nocebo story—I definitely choose the first. Luckily, I've seen enough examples of healing done by alternative methods to grant a healer some real authority.

Remember that feeling impacts the speed and certainty with which story becomes experience. In this situation both fear and relief are strong feelings. I far prefer the relief of not being a candidate for Alzheimer's to the fear that I will live the nightmare that was the last twenty-two years of my mother's life.

Could changing my story change the outcome of a toss of the genetic dice? Why not? I find myself now deeply interested in those identical twins who didn't both get Alzheimer's—did one have a personal story that was more positive, upbeat and optimistic than the other's?

When I hear of women who have double mastectomies because there is breast cancer in their families, I deeply empathize. I understand. Their Big Story is thoroughly supported in our world. I realize how difficult it is to revise it for themselves, and how much they are willing to do in the physical realm because they don't understand the power of their own consciousness. Would I tell them not to do it? Of course not. But at the same time I would hope that they could latch onto the possibility of changing their *story* instead.

We have many Big Stories. We have at least one about everything that is really important to us—all the Big Issues in our lives. It has been said that there are four foundation blocks of a fulfilled life—*Health, Wealth, Love and Self-expression.* These tend to be the general topics of our Big Stories.

These stories are compounded of what we have been told, what we have experienced, our interpretations of our experiences and the meaning we have ascribed to them, and the "truths" and "facts" of consensus reality. We believe them, and we tend to repeat them to ourselves, to our family and friends, to strangers on an airplane, to anyone who wants to get to know us, anyone we want to get to know. These Big Stories are our reality. We are living them out day after day.

It's important to remember that however big they seem, however long they've been around, they are still stories and can be changed. Looking at those big issues of health, wealth, love and self-expression we can discover what the stories are that we are telling about them, and how they make us feel.

What are the stories that have been handed down in your family as truth? And how do you feel about that truth? A dear friend of my husband's and mine spent his early adulthood both attempting to get around his family story and preparing for it. All the males in his line had developed heart problems. All had died before they reached the age of fifty. So he watched his cholesterol and his triglycerides, did his best to stay fit, and bought a lot of life insurance. When he was 49 and in apparently good health he came out of a concert one night and stepped off the curb directly into the path of an oncoming car. He was killed instantly.

So strong was his constantly repeated and thoroughly internalized Big Story about the males in his family not making it to fifty that, blocked from working out one way, it found a different route. He paid intense attention to that story, and even though his attention was focused on keeping it *out* of his own experience, his attention to it and his belief in it, brought it about. "Lo, the thing I have greatly feared has come upon me." (Job, 3:25)

Noticing the Story Running in the Background

Recently a friend called me to see if I could help her cheer up—she was feeling, she said, "out of alignment with the universe." That was a very good way to describe her bad feelings, because to be *in* alignment with the universe means (again, trust me on this for the moment) that you are in alignment with well-being.

Sonja was in the midst of a wanted/not wanted divorce. She had met someone she hoped might be a part of her future. A flirtation had begun between them, and then she discovered that the man was married. "I pick duds!" she told me. "Bad guys. Guys who will abandon me just like my father did. No matter how I try to find somebody better, I pick a bad guy every single time."

What she had just told me was a Big Story about Love that she had been telling herself over and over, probably since she had first felt abandoned by her father. Having told it over and over, she was living it—as is the way with stories. Sonja and I had talked about Story Principle. But at that particular moment it wouldn't have been useful for me to remind her that she was the author of that negative story and could change it whenever she wanted. Her negative feelings were giving her the message that a story needed changing, but right now she just wanted those feelings to go away. It was much more important for her to find a way to begin to feel better than to consider how her uncomfortable reality had been created.

I suggested to her that she was "between stories," and that such a transitional place was always uncomfortable. Her marriage wasn't fully over (she hadn't quite let go of wishing it didn't have to be), and she didn't have a new relationship to put in its place. I suggested that she do a "Quantum Meditation" (in the Putting it to Work section following this chapter) because what she needed right then was to get "out of herself and her story" for a while, to disconnect from her pain a bit so she could move through this transitional time.

The Quantum Meditation allows a direct connection with the power and freedom of our author-self. Any time we make that connection, it helps us feel better! I hoped that when she felt better she might be able to hear and accept the idea that her experience had been an out-picturing of the Big Story she's been telling that she can only find bad guys. Better feeling could allow her to summon the energy to change that old story, so that she won't go on fulfilling it over and over.

It can't be said too often. Any story—*no matter how old or often repeated*—can be changed at any time. But the first step is to stop telling it.

[1]Tolan, Stephanie. "The Dying of the Light," *More, Smart Talk for Smart Women,* October, 1998, 152-155.

Putting it to Work

Discovering your Big Stories

Get four sheets of notebook paper or divide one into four equal sections. Give each a heading: *Health, Wealth, Love* and *Self-Expression.* You can do this on a computer if you'd rather, taking each of the four subjects in turn. Under each heading write down what your life has been like in regard to this subject. This doesn't need to be in the form of a "story." You can just record notes about what your experience has been in these four areas of your life.

Under *Health* you don't need to include everything you'd record on one of those long Medical History forms they give you when you begin seeing a new doctor. Just put down the major health issues you've encountered and whether your experience has given you a sense of being unusually healthy, unusually sickly, or something in between. Do you find yourself worrying about your health a lot? Are there genetic predispositions in your family that either have or have not shown up in your experience? Anything you can think of about your health and how you feel about it, including your weight, your general level of physical fitness, whether you are on medications for ongoing conditions, can go here.

Under *Wealth* you might want to consider first what that term means to you. For some it might be exclusively a money issue—for others it includes quality of life, such things as living where you want to live, having time to engage in activities you like, doing a job you really enjoy. Once you've defined the term for yourself, record, as you did under *Health,* what your life has been like on this subject. What do you worry about? What are you grateful for? What do you hope for in the future? How does this whole subject make you feel—cheerful, frightened, confident, frustrated, discouraged, angry, triumphant?

Under *Love* consider all the relationships in your life—your family of origin, your friendships, your romantic experiences, partner[s], spouse[s], your children, even your pets. Here again, note your overall feeling response to the whole subject of love in your life and also to each aspect of it, since they may be very different. Which have "worked" for you and which haven't? Which have given you the greatest satisfactions and which the least?

Under *Self-Expression,* think about what you most like to do, what seems most clearly connected to who you feel yourself to be (there may be one thing or several, and they may be genuine passions or just activities that you enjoy) and how much time you've spent doing it or them. Have you pursued one or more as a profession? A hobby? An occasional pastime? Also consider what you've achieved—what has been the outcome of your activities in these areas. And how have you felt about that?

Take your time with this process—perhaps focusing on one area one day and another the next. When you've recorded everything you want to record on each of the subjects, you'll begin to see patterns in them. You may also find a pattern that ties two or more or even all of them together. If you're sick a lot you may have to spend a great deal of money on insurance, doctors, treatments and medications. What you most love to do and would like to do for a living may not seem to provide a living wage, and that could have an impact on those you love.

Having created these notes as your life experiences, consider them as "story." As experiences they have been real and probably feel quite real right now. As story they can be changed. Any aspect of these four building blocks to a fulfilled life that makes you feel more negative than positive is a focal point for changing your story. Remember that experience creates belief while a changed story needs *a suspension of disbelief.* No future story has to be limited by what has happened in the past. If we couldn't move beyond what has already happened in our lives, humanity would still be living in caves.

Once you can see the aspects of past experience that you would like to see changed, you can begin to engage your story telling abilities to change them.

Quantum Meditation

Quantum physics has shown us that the building blocks of the universe aren't blocks at all. Looking for the smallest bits of matter from which the material universe is made brought us the startling evidence that it isn't made of matter at all, but of bits (pulses or vibrations) of energy and information. Picture a montage of images from the very large to the very small, from galaxies to stars to our own solar system, to the earth to a mountain to a tree, to a branch, a leaf, a cell, a molecule, an atom, a sub-atomic particle. When we go far enough downward, focusing on smaller and smaller fields, we finally and surprisingly find ourselves not confined to an unimaginably tiny place, but loosed into an unimaginably big one, the field of all possibilities.

The Quantum Meditation is just such an exercise, but undertaken through the focus point of our own bodies. We begin in the realm of our material selves, move gradually down to the very small and out again into the infinite reaches of consciousness where our author self is connected to Cosmic Imagination.

This meditation or one like it is a way to connect with our nonmaterial being, to break for a time the chains of material limitation, to expand our minds and our hearts and rest in a place of infinite power.

You can use these words or write a script for yourself creating images for yourself. If you have a Fellowship of the Story or a Story Partner, you can take turns being the guide. If you are working with Story Principle on your own, you can record yourself and use ear phones to listen to your own voice guiding you through it. Whether you are reading the meditation script for someone else or recording it for yourself, keep your voice calm and gentle, pausing for a few seconds whenever you come to an ellipsis (. . .) in the script.

Instructions for doing the meditation are pretty standard. Settle yourself comfortably (sitting up rather than lying down so that you aren't as likely to go to sleep) in a place where you won't be interrupted. If you can sit in yogi position with your legs crossed, that's fine. If not, sit in a chair with your spine straight and both feet on the floor. Breathe deeply and feel yourself relaxing and letting go of your immediate outward concerns. Close your eyes and listen to the guide or to the recording.

Imagine yourself in a place of great beauty and peace. It may be a place in nature—a green grotto with a splashing waterfall, a deserted beach or a silent mountaintop . . . It may be a candle-lit cathedral or a book-lined study with a roaring fire in a stone fireplace. . . . Wherever you are, know that you are totally safe. Sit for a moment absorbing the peace and safety, aware of whatever sounds, smells and sensations you can imagine in this place. . .

When you are fully there, fully relaxed, move your attention to your body. Notice the feeling of your clothing against your skin, the surface under you, the slightest movement of air against your face . . . Focus on these physical sensations, letting go of any awareness beyond your body . . . Notice your breath moving through your nostrils . . .

See if you can become so still that you begin to feel the movement that comes from the beating of your heart. . . . Imagine the movement of your blood cells through your veins and arteries, flowing forward with each pulsing of your heart . . .

Now visualize the blood cells as they move, and focus your attention on one of them. It is a smooth disk with a round indentation in the center, a rich red in color. Watch it drift in the stream of plasma, bumping into others like itself. . .

Now take your attention closer, focusing on the outer membrane of the cell. You find you can see the molecules of the cell membrane, like soft beads of varying shapes and translucent colors . . . Imagine yourself slipping easily between them and into the center of the cell, moving down among the bright red molecules of hemoglobin that give the cell its color . . .

After a moment, you slip into a single hemoglobin molecule. You find that it is like a mobile, mostly space with wide spiraling ribbons of bright red and hexagonal grids—Float there for a moment . . .

Now you feel yourself drifting into an atom of iron. There is no sense of size or substance here—it is as if you are floating into our own solar system. In the distance you see a bright center sphere around which spin dots of light. You watch them move, circling constantly, as you drift among them. . . You feel yourself being pulled inward toward the bright light of the center. As you get closer the light gets brighter, but it doesn't hurt your eyes. . . Now, gently, you are sucked into the center of the light and feel its soft warmth. . . There is nothing now except the light, at once white and iridescent, its rainbow colors dancing around you. You feel that you have left the realm of physical reality behind and have entered the nonmaterial realm of consciousness, of imagination itself, held by light, aware of infinite space, infinite power, infinite possibility . . . Time has no meaning here. Seconds expand so that you may stay as long as you wish, feeling yourself becoming that light, that space, that power. . . .

When you are ready, allow yourself to drift back through the atom . . . into the molecule . . . through the membrane of the cell and into your bloodstream, returning to the feel of your body, your heart beating, your breath moving in and out, the air against your skin . . . Gently and slowly allow yourself to return to your normal consciousness, taking a few long, slow breaths and—when you are ready—open your eyes.

This meditation can take as little as five minutes or as long as half an hour and can be done as often as you wish. You are likely to find, if this meditation speaks to you, that you return to your normal consciousness with a greater sense of peace, power and possibility.

8
Foundational Stories

"Whatever is true, whatever is noble, whatever is right whatever is pure, whatever is lovely, whatever is admirable, if anything is excellent or praiseworthy, think about such things."

–Philippians 4-8

Foundational Stories are the trickiest of the threads that make up the story of our lives because they're the hardest to recognize. They exist on a level of awareness most would call subconscious, where they affect both our Big Stories and our Small Stories, and—of course—our lives, most often without our knowing it. Like all stories, they can be positive or negative. If we don't know that we're telling them or what they are, we can't determine how they are shaping or distorting our other stories and whether they need changing.

Foundational Stories tend to establish themselves when we are very young. They get put into place by our earliest experiences and the teachings of the adults in our world.

Human infants have much to learn in order to navigate the world, and adults believe they have the information babies, toddlers and children need. It doesn't occur to very many adults that what they are teaching their children (or other people's children) is story. They believe they are teaching them truth. Reality. The rules. And so, of course, that's how they present their stories.

For the most part children don't have the tools to discern which of the things they are being told will be good for them and useful in their lives, and which will be harmful. They tend to absorb them all. The Foundational Stories parents teach their children are likely to be many of the same Foundational Stories *they* were taught. After all, those stories were taught (and experienced) as truth and most of them have been around a long, time. These may be stories about family culture, about politics, ethnicity, religion, etc. They are about "our place in the world" or "the people who are out to get us" or "principles that can be depended upon."

Some Foundational Stories grow out of the parents' particular time, place and experiences. People of my generation tend to have *particular* trouble with scarcity thinking, I've noticed. Our parents came of age during the Great Depression. However they managed to survive that time, they wanted their children to be protected from such deprivation. So we were raised with a constant litany of scarcity-based proverbs: "Don't count your chickens before they're hatched;" "A bird in the hand is worth two in the bush;" "A penny saved is a penny earned;" "Save your money for a rainy day," and so on.

Because we were given these rules to live by in a world that believed (and still does) that the planet's resources are limited and in danger of running out, it is no wonder that a Foundational Story of financial uncertainty and a fear of lack and scarcity got formed and steadily reinforced.

Trauma

If a child is born into a world where no one is available to give her the love, attention and touch she needs, she may "fail to thrive," even if her basic physical needs are being met. Such a child, if she makes it into adulthood, is likely to live her life with a Foundational Story that the world is a dangerous place in which survival is always in question. When this neglect happens in infancy and lasts through the early years, the Foundational Story is likely to be highly negative. *World: dangerous. People: unreliable. Self: not worth noticing.*

Such a response to extreme neglect is very common; however it is not inevitable. There are some kids who seem to be born with an innate resilience that protects them from being so damaged. Such children somehow are able to establish for themselves a strong, positive Foundational Story in spite of their experience. *World: often dangerous. People: mostly unreliable. Self: Okay anyway. Or Self: able to handle anything.* Whatever accounts for this fundamental resilience, it seems both wonderful and rare.

Child's Eye View

The trauma that can create a negative Foundational Story may be physical (injury, disease, abuse) or emotional/psychological (apparent parental disinterest, extreme parental anxiety, or hostility, competition, and constant belittling from parents, siblings or caregivers). It's important to remember that it isn't only what happens to a child that creates trauma, but how the child interprets it. An extremely sensitive child may take normal sibling rough and tumble as hostility, and competition as a purposeful effort to hurt her.

Kids who misunderstand a painful situation may turn it into a fully traumatic one. If a very young child who naturally perceives his parents as all-powerful develops a serious physical illness, he may not understand why his parents don't "fix" it. Even though he may eventually come to understand that there was nothing they could do, a Foundational Story that the people he depends on are not reliable may have been established and may remain in place because it isn't conscious. *World: dangerous. People: unreliable or weak. Self: broken or damaged, and alone.*

A child who gets mostly negative attention from parents and caregivers is likely to form a Foundational Story of unworthiness. A child whose personality conflicts with the personalities of parents and siblings may feel permanently out of place. *World: generally unwelcoming. People: critical and condemning. Self: unworthy and out of sync.*

Parents who actively both show and express their love, on the other hand, help their child form an original Foundational Story of worthiness and value. Well established by the time the child goes to school, this story may hold up in spite of negative or even traumatic interactions with teachers or other adults and other children. *World: Welcoming and interesting. People: Not bad. Self: strong, good, worthwhile.*

Foundational Stories are silent background, mostly unquestioned because unnoticed. It can take real effort to uncover them and decide whether their effect on our current experience is helpful to us or not.

Julia Cameron, in her book *The Artist's Way,*[1] provides some excellent methods of discovering Foundational Stories. One exercise she suggests is to write down positive affirmations about ourselves or some personal project we care about, listening as we do so and recording what she calls "blurts." These are the arguments that the voices in our heads throw up against our positive affirmations. Follow the trail of the blurts and we may find a hidden Foundational Story. Sometimes we also discover the source of that story, as the voice may seem to be the voice of the person who first told us the story.

Using this exercise I discovered a Foundational Story (a specific instance of the more general scarcity story) about the place of work and of art in the world. The larger story was that adults had a responsibility to work (work *hard!*) to earn enough money to support their families. I come from a family of avocational artists, so art was important. However, art was fun, not work—work was by definition *not fun.* Furthermore, art didn't pay well. It couldn't support a family. It was okay, even admirable, to do art, but only as a hobby. Anything else risked poverty or catastrophe and wasn't fair to others depending on you.

When I discovered the Foundational Story that art couldn't support a family, my husband and I had been supporting our family through art (he as a theatre director, I as a writer) for about 20 years!

But here's the effect that story was having on our experience: we were often on the brink, often struggled to make sure the mortgage got paid (which in those days it occasionally didn't). Much of the time, accurately or not, I felt real poverty was imminent. Besides that, since I was doing something that was fun rather than work, I did it in such a way as to be sure it was as little fun as possible. I worked *hard* at it. I spent many hours at it, whether the writing was going well or not. I seldom allowed myself to just blow off a day if the writing wasn't going well—I forced myself to write anyway. I was constantly anxious about whether my books would sell *and expected them not to.* The more I worried about book sales, the harder I worked at writing the next book. I was practically allergic to play and had virtually lost the ability to have fun.

This Foundational Story affected my Big Story about Wealth (I couldn't have it), and also about Self-Expression (my favored kind was irresponsible).

Here's the amazing thing. Our most damaging Foundational Stories, when discovered, may reveal themselves immediately as false. And just that quickly, they can be over-ridden! They have remained, throwing up obstacles and thwarting our real wishes, only because we didn't know what they were or even that they were there.

That was the case with this one about art and work. Once I discovered it, I saw that it couldn't hold up against logic or against our actual experience. Not only had our family been supported by art for years, but I knew plenty of artists who were doing quite well financially! Besides that, there was Joseph Campbell, a genuine cultural authority, telling the world that it was a good, even *noble* thing to "follow your bliss." As soon as I discovered the story, its negative effects began to dissipate.

There are other Foundational Stories that are harder to over-ride, even if we discover what they are. Religions are really good at providing lasting ones. How many people have you heard calling themselves "recovering" Catholics or Southern Baptists, for instance? If you have been inculcated with the idea of original sin, told by those in authority over you that you are by your nature somehow evil—sons of Adam, daughters of Eve—and constantly in danger of breaking a rule that will send you to hell for eternity, your Foundational Story is *unlikely* to be one of healthy self-love.

School teachers who humiliated us, put us down, shamed us, helped many of us create negative Foundational Stories about ourselves as well—too talkative, too arrogant, too intense, too sensitive. Or not smart enough, not good at what we needed to do, troublesome. Today's environment of constant testing in school may give some children a foundational message that everything they do will be judged and evaluated, that they are in danger in every moment of being wrong—giving wrong answers, making wrong choices. And that those wrong choices will close off their future hopes and dreams.

Since most of our Foundational Stories were established when we were very young, they've been with us, helping to create a reality that supports them, for a long time. Even though some can vanish like mist in the sunshine when we discover what they are, more often it takes concerted effort to find and revise them. They form our beliefs about who we are, how we fit into the world and what that world is like.

Creating more positive ones is no harder than creating any other story, but we may have to marshal a lot of evidence to support them and repeat the new ones over and over, because the old ones feel so real and are so easy to fall back into.

When we slip back into an old story in our lives, the good thing is that we just have to remind ourselves of the new one and continue moving ahead. The basic principles of our new Foundational Stories can be very broad and general statements of a few words that can accompany us through our days, to help us remember our new direction. I have several such affirmations that I find myself repeating over and over on particularly trying days. The first was just there in my head one morning when I woke up: "All is well, all is well, all manner of things will be well." It felt as if I were "hearing" the words rather than thinking them, and it didn't sound like my own voice. I now know it as the voice of my author self. That statement works to pull me out of the Foundational Story that I carried through all my years of depression: "What can go wrong will go wrong."

When I find myself feeling unworthy of some good thing in my life, I repeat another saying that also appeared in my head as I woke one morning: "You are infinitely loveable and infinitely loved." When I first heard those words, my eyes filled with tears. The rush of positive emotion reminded me how powerful the effect of a few words can be. I understood, suddenly, how my mother's response to misbehavior of any kind when I was a child—"You ought to be ashamed of yourself!"—could have set a Foundational Story of unworthiness so deeply into my consciousness.

If I had known then that I was an aspect of the oneness, that I was "made of light," I might have recognized that the negative feelings my mother's words created in me were proof that they didn't reflect my deepest self and so were not a useful story. Instead, I assumed that those negative feelings meant that her words were truth, so I built them into the foundation of my life. That is a story I still work at over-writing.

As we focus on finding and changing negative Foundational Stories, it's important to direct our attention away from the cacophony of negative stories about the state of the world and "man's inhumanity" beamed at us steadily by the media or repeated by the people around us, and pay more attention to the wonders of our planet—the flower growing up through a crack in the pavement, a "shooting star" in the night sky, a full moon rising over the trees, birds greeting the dawn.

World: pretty darn good. We can remind ourselves of the many people organizing to help other people, to help animals, to work on global issues like peace and justice and health, and of the kindnesses that have been shown to us personally. *People: mostly pretty nice.* And we can remember and value the kind, generous things we've done, the love we've shown others. *Self: Definitely okay, moving toward terrific.*

[1]Cameron, Julia. *The Artist's Way, A Spiritual Path to Higher Creativity.* New York: G.P. Putnam's Sons, 1992.

Putting it to Work

Revising a Foundational Story

Finding Negative Foundational Stories

If you did the exercise about your Big Stories, you may have uncovered clues to some deeper Foundational Stories in the process, whether you were aware of it or not. Go back to what you wrote about *Health, Wealth, Love* and *Self-Expression,* looking for patterns relating to the three major subjects of Foundational Stories: *World, People, Self.*

On another sheet of paper write down those three headings and jot down under each whatever comes to you as you consider what you wrote in the Big Story exercise. Is there a sense throughout what you wrote for the exercise that large, often unseen and unrecognizable forces may work against what you want or need in your life? Do you expect to have to struggle to accomplish anything? Do you see there an expectation that other people can't be relied on, may be out to get you, are more likely to treat you badly than to treat you well? Are there threads running through your Big Stories that show distrust of yourself or a feeling of unworthiness?

Whether you find these kinds of negative threads running through your Big Story exercise or not, take a look at the three headings of *World, People* and *Self,* and see whether you get a mostly positive feeling or a mostly negative feeling about each of them.

Now try writing down a positive affirmation under each of them in turn and see what "blurts," as Julia Cameron calls them, come up. (*The Artist's Way,* pp. 34-36) What arguments against the affirmation pop into your mind? If you wrote, for instance, "I live in a beautiful, joyful, abundant world," and you are immediately assailed by thoughts of pollution, global warming, loss of species and natural habitat, you have found a story (a major story in the collective consciousness just now) that the world is in serious trouble. Try writing affirmations about the planet's resilience and ability to heal itself, about the work humans are doing to alleviate the problems we have created, and see what comes up. The arguments your mind throws up to these give you the negative stories you want to change. Perhaps the story working under your attitude about the world is that it is fragile, vulnerable and limited. This is the story you wish to change.

The arguments you found against the affirmation about beauty, joy and abundance provide a clue not only to your story about the world, but to your story about people. Pollution, loss of habitat, and global warming are all problems created for the most part by human beings. Under *People* you might want to write an affirmation about humanity's creativity and ability to undo what they have done and see what blurts appear in response. Try an affirmation that the people in your life are trustworthy, loving, dependable.

As you work with your stories about *Self*, go back to what you wrote when you were creating a hero self you could believe in. If there were positive attributes you would have liked to write down about yourself but couldn't get yourself to believe, consider them again, write them as affirmations under *Self*, listen to the blurts, and see what you find.

The more affirmations you write under these three headings and the more blurts you encounter in response, the clearer you will become about the Foundational Stories that form your mental atmosphere.

Marshaling evidence against a negative story

Once you've tracked down a negative Foundational Story you want to work with, put two pieces of lined paper next to each other. Don't do this exercise on a computer, because you're going to want to have paper to work with at the end of the exercise as a symbol of change. On one page write out the negative story you have found that you're living with. On the other, write all the evidence you can marshal against that story.

If you have found negative stories about the condition of our world, you can search out stories about nature's ability to heal itself. Here's one small example: I live in a woods where I noticed one day what I call a "linear grove" of sweet gum trees. Many years ago a huge sweet gum was blown down, pulling up a root ball nearly seven feet tall. Its lower roots remained in the ground when it fell. Since then what had been limbs on the upper side of the blow-down have turned themselves into sweet gum trees as tall and as thick as most of the trees around them. The only way to distinguish this straight line of sweet gums from the others is to notice that they aren't growing out of the ground, but out of the trunk of what had been their "parent" tree. That is only one of many examples I have found in my daily walks in nature of the resilience and determination of life. You can also do some research on what humans are doing to alleviate environmental problems we have created.

If under *People* you have found a Foundational Story that humans are violent and aggressive and peace seems to get farther and farther from possibility, look for the abundant evidence that we are less willing to accept war as a method of handling differences than in the past. Check out the peace movements, peace programs at colleges and universities, or the Human Security Report (http://www.humansecurityreport.info) that documents "a dramatic, but largely unknown, decline in the number of wars, genocides and human rights abuses over the past decade."

If you have been getting most of your information about the world and humanity from the popular media (news broadcasts, newspapers, television and movies), most of the evidence you've been collecting will have supported negative Foundational Stories. It's important to show yourself that the media's preference for fear stories (both the fictional ones in dramas and the ones that make up the news) is a distortion of human experience. There is actually far more evidence to support positive stories than negative ones once you begin looking, though you may have to be creative about where you look! One place you can begin is the "good news" page of the website of the Foundation for a Better Life, http://www.values.com/good-news.

Meantime, keep in mind the fundamental point of Story Principle that *everything* is story—either currently existing in our experience or on its way to becoming our experience. No matter what predictions scientists are making, those predictions are stories. Get in the habit of challenging negative predictions no matter where they come from. This is not to deny them—this is to tell yourself a different story based on different evidence and a different perspective. For instance, in an article in the June-August 2006 issue of *Shift,* the journal of the Institute of Noetic Sciences, evolution biologist Elisabet Sahtouris gives "Seven Reasons Why I Remain an Optimist." In her response to global warming she doesn't suggest that the earth's climate is *not* changing—but she focuses on the opportunities the change presents rather than the difficulties.

Negative stories about yourself (*too loud, too intense, too sensitive, not smart enough*) can be offset by finding times in your life when a negatively labeled attribute actually allowed you to accomplish something you couldn't have done without it, or occasions when you accomplished something in spite of it. (It was a major revelation to me to realize that the trait my family assessed as "too emotional" was a critical gift for me as a novelist, as was my extreme sensitivity to my own pain and the pain of others.) Take this opportunity once again to look back over your life with new eyes and take credit for all of your accomplishments, whether your Foundational Stories have recognized them or not.

If your religious training gave you negative messages—original sin and unworthiness, for instance—you may be able to use the same scripture that supported those messages to contradict them. If you incorporated an image of a judging, vengeful God you can find other images of a loving, forgiving, supportive one. The world's scriptures provide an astonishing and paradoxical array of stories, parables, tenets, principles, beliefs and rules. Where you find contradictions there is nothing wrong with choosing the most positive interpretation.

Having found as much positive evidence as possible to counteract your old negative stories, create for yourself a ceremony of release for those stories. You may choose to burn the pages on which you've recorded them (throwing them into a fireplace, a campfire or a "burning bowl") or you may prefer to tear them up or shred them. To make the message particularly clear, you may want to tear or shred them, *then* burn them and finally *bury the ashes!* What you are doing is "documenting" the destruction of the old foundation so that you can accept beginning again on a more positive one, more fully aligned with your own well-being and the structure of the universe. (Which we'll get to in the next chapter!)

Keep the pages of evidence you've collected for more positive stories to reread as often as necessary and to share with others.

Create—or find—some positive affirmations that speak to you.

The best ones may come to you, once you begin looking for them, as mine did to me, unbidden. They might pop into your mind or appear in your dreams. The good thing about these is that they come from deep inside yourself, so they are crafted to fit your needs and part of you is already in agreement with them. A song might be running in your mind when you wake, reminding you to "accentuate the positive." Or you might see an affirmation on a bumper sticker or a billboard that zings you and sticks in your mind. A workshop leader told of seeing a billboard advertising a credit card that gave her a positive mantra she'd been using every since: "You are pre-approved!"

There are plenty of sources of positive affirmations you can find online or at libraries and bookstores. The important thing is to find statements that really connect with you, that resonate and make you feel so good that you don't immediately begin arguing with them. These are the statements you will want to tape up to your mirror, keep in your wallet or purse, stick up on the refrigerator with magnets, keep in your car. The more you see them, the more you say them, the firmer their rooting in your consciousness, and the more evidence of them will enter your experience.

9
Friendly Universe?

I think the most important question facing humanity is, "Is the universe a friendly place?" This is the first and most basic question all people must answer for themselves.

–Albert Einstein

Our whole lives are flavored by our answer to this question—it could be said to be *the* critical Foundational Story. And while you can find websites all over the internet trying to answer that question for all of us, no one can answer it for anyone else.

These days my answer to Einstein's question is yes. The universe I live in is friendly. More than that, its essence—energy—substance—is wholly positive. I think of its essence as *light* (or *goodness* or *love*). Its infinite diversity is a manifestation of its infinite possibilities. But each individual expression of that diversity, however separate it appears, however negative it may seem, is an aspect of a unified positive whole, just as every color in the spectrum is an aspect of white light.

At the level of story we see both light and darkness, which we tend to think of as opposing forces. *But darkness is not a force in itself.* It is simply an absence of light. Light can be obstructed. We can turn away from it. We can close our eyes to it. But darkness does not, cannot, invade and obliterate light. There cannot *be* a "complete extinction of light by darkness." Substitute "good" for light and "evil" for darkness and the story remains the same.

Are you saying that evil doesn't really exist?

Yes and no. Evil is not a *force* inherent in the structure of the Universe. Just as darkness is not a force that can invade and obliterate light, there is no evil force that is capable of invading and obliterating goodness. There is no duality or polarity, no struggle between opposing forces, in the essence and structure of all that is, the creative level from which story emerges—the level of Cosmic Imagination where our author self functions.

But on the level of story where we exist as character, evil clearly exists. Story level is a level of contrasts, of apparent duality and polarity. Evil is one of the ingredients for creating particular kinds of story. And evil has been incorporated into many of the stories people are and have been living out.

In my book *Flight of the Raven* there is a character, Charles Landis, who is the head of a terrorist group. He is—without question—a villain, and what he wishes to do is readily identifiable as evil. There is also the character of Elijah, the boy who is kidnapped in the early pages and held by the group of terrorists Landis leads. It is Elijah's experiences at the hands of the terrorists that allow him, *require* him to discover his own internal power.

Within the story Charles Landis is the bad guy and Elijah is the good guy, but if you were to ask me, as author, which is the *better* character, I couldn't answer. At the level of author the question has no meaning. Both are essential to the story. It is the pressure from the bad guy that brings out the strength of the good guy. In a real way, the villain helps create the hero! And certainly their interactions create the story.

A friendly universe is a fundamental aspect of Story Principle. The essence of the universe is positive, which means there is literally nothing too good to be true. *Good is ultimate truth.* What that means is that the appearance and experience of evil within the stories we are telling and living, has no reality *beyond the level of story* and no lasting effect.

Kierkegarde has been quoted as saying, "Life can only be understood backwards, but it must be lived forwards." When something happens to us that we don't like, we label it bad. When someone purposely hurts us or someone else, we label that person or that act evil. But even within the story, once we have moved beyond the immediate pain, what grows out of the experience can have a positive impact on the rest of our lives, if we are able to let go of our original judgment. What we called bad, even evil, when it happened (as we lived it forwards) can turn out to be among our best or most important experiences when we look back later.

For that transformation to happen, however, we must shift our attention and allow ourselves to see the good that has come of it. We're not always successful. If we focus on the negative and continue to keep the story of the awful thing that happened or the evil person who hurt us alive in our memory and our feelings, we can assure that the long term impact remains as bad as we thought or—more likely—grows even worse. What we pay attention to expands in our experience. This is the reason that forgiveness is so critical to changing our perspective on evil. We do not forgive in order to deny our pain or let the perpetrator of evil escape the consequences within the story, but to release our attachment, to let go of the resentment, the memory, the blame that keeps us tied to the negative experience and keeps it or others like it repeating in our lives. Forgiveness allows us to move through and past the pain and focus our attention on whatever positive outcome (even if it were only discovering our ability to forgive) we can find.

When bad experience spurs us to acquire new coping strategies, to grow stronger, to focus more completely on the positives in our lives, it turns from blight to blessing no matter how it felt at the time. It all depends on the story we choose to tell ourselves about it.

But what if the evil experience results in our death—or the death of someone we love?

Death is a story itself, such a profound one that it deserves a chapter or two of its own. For now I will say that as with every other aspect of Story Principle, the story we tell about death creates our experiences around it. The existence of death at the level of story does not contradict the idea of a friendly universe or invalidate the statement that good is the ultimate truth. Radical as that statement may seem, adopting it as the foundation of every story we tell is an important step to living a more peaceful, more positive life.

No pessimist ever discovered the secret of the stars, or sailed to an uncharted land, or opened a new doorway for the human spirit.

–Helen Keller

Martin Seligman, the psychologist from the University of Pennsylvania who is considered the "father of positive psychology," said in an interview with *Omni Magazine*[1] that optimists have many advantages over pessimists—they do better in sports and at selling insurance, "they're more resistant to infectious illness and are better at fending off chronic diseases of middle age." He went on to say, "Optimists have a set of self-serving illusions that enable them to maintain good cheer and health in a universe essentially indifferent to their welfare."

For Seligman (at least at the time he gave this interview) a universe essentially indifferent to our welfare was *reality*. In the face of that reality, an optimist's good cheer and health surely had to be *illusion*. Never mind that the positive effects of that illusion could be documented.

Story Principle simplifies things considerably by removing the illusion/reality distinction. Illusion and reality are not either/or, but both/and. Having identified everything in our experience as story, either real in this moment or on the way to becoming real, we can readily see why an optimist's story leads to health and success. Optimism (living with a positive story) *is* self-serving. The question is whether we choose to cultivate it and put it to work for us. How do we answer Einstein's question?

Living in a random or indifferent universe makes it hard to be an optimist. Living in a hostile universe would seem to make optimism impossible. That leaves only one genuinely self-serving answer to Einstein's question.

The Power of the Positive

Within a universe in which the unified field, the central force, is friendly, positive—light—every positive story is in alignment with the universe itself. Every negative story, then, is out of alignment. Every positive story has the power of the universe going for it, supporting it, activating it, while every negative story resists the power, obstructs the light, creating shade, shadow, darkness. A story with a positive "charge" brings with it, then, considerably more energy in its support than a story with a negative charge, though both will manifest as actual experience when held consistently in consciousness.

Becoming Aware of Feeling Bad

We can readily tell when our story is negatively charged and out of alignment with the universe, because a negative story *makes us feel bad.* When we become aware that we are feeling sadness, depression, jealousy, frustration, grief, anger, resentment, or fear, in any of their various intensities and permutations, instead of assuming that those feelings are the inevitable result of some negative reality, we can recognize them as warnings that the story we are telling ourselves in the moment is negative and likely to bring us more of the same. We can then immediately look for a way to change our story.

The trouble is, we are not always clear about what we are feeling, or why. In 1854 Thoreau wrote "The mass of men lead lives of quiet desperation."[2] Most of us not only believe that, but assume there is little we can do to change it.

The culturally accepted story that external forces affect us in ways we can't control gives us the sense that feeling bad is inevitable much of the time. Of course we feel bad when something happens that we don't like, we say. There's no way to avoid it. We may even think that people who seldom feel bad (even when bad things are going on in their lives and in the world) are either a tad loony or not quite bright. This is what gives Pollyanna a bad name. There's a bumper sticker popular with progressives at the moment that says "If you aren't outraged, you aren't paying attention!"

I remember very clearly telling a friend of mine about twenty years ago that a group of fundamentalist Christians who were cheerily preparing for The Rapture could be recognized as utterly wrong-headed by the very fact that they were so bloody cheerful. "Anyone who can have that attitude in this world," I said, "is a fool to begin with!" Outrage, fear, sadness, frustration—none of these feelings that pervaded my life at that time could have spurred me to question the story I was accepting as reality. I simply assumed that intelligent people couldn't help but feel that way.

Even worse than accepting bad feelings as inevitable is losing our awareness of them altogether. We may live with negative emotions so much of the time that we adjust to them. They feel normal. They come to seem like who we are. Unless those feelings cross the line into major anxiety attacks, clinical depression, suicidal impulses or uncontrollable rage, we simply don't notice them. So we go on telling or accepting from the world around us the stories that cause those feelings and keep those stories operating in our experience.

How Often Do You Feel *Good*?

Instead of becoming aware that we are feeling bad, therefore, some of us might find it easier to check on whether the stories we are telling or accepting are serving our best interests by asking ourselves how often we feel good! When was the last time we spent the *majority* of our day feeling appreciation, happiness, pleasure, love, joy, gratitude or bliss? If we can't remember a day like that, if those feelings are rare moments of light in the shadows of our normal experience, it is time to look carefully at our stories and see what we can do to change them.

It is commonly thought that to feel good we must have good things happen to us. As we learn to put Story Principle to work in our lives, we discover that if we want to have good things happen to us, we need to find stories that allow us to feel good.

I was fashionably nihilistic in college (not surprising, given the fact that I was so often severely depressed—nihilism seemed not only fashionable, but logical). Gradually, over the years, I managed to work myself up to being merely pessimistic, with recurring bouts of major clinical depression. Though my depressive episodes lessened as I explored metaphysics, I wasn't really able to change my answer to Einstein's question until fairly recently. I kept getting caught in the stories being acted out all around me, the shades and shadows, good and evil, of the Saga. It isn't always easy to hold onto my new answer because the stories of a random or hostile universe are not only the stories I used to believe, but go on being told all around me by people who are considered or consider themselves *authorities.* The media are constantly with us, feeding us fear-based stories to keep us tuning in. Meantime, we see the evidence of people living out difficult and painful experiences that are the result of fear-based stories.

Be aware that the fashion of the times, like the fashion for existential depression during my college years, is for cynical, negative, fearful stories that make a cheerful, joyful, optimistic person stand out, not necessarily in a comfortable way. Pollyanna *does* have a bad name. It can be challenging to risk being seen as out of touch with "reality." But we can remind ourselves of the concept of placebo and nocebo ideas.

In their books about belief both Benson and Lipton tell stories of patients whose acceptance of a frightening, negative medical diagnosis given to them by a respected authority and supported by family and friends literally killed them. We inhabit a psychological environment pervaded by nocebos so we need to learn how to tell and maintain a positive story while surrounded by those who believe otherwise. It takes a steady commitment to our own well-being and a willingness to pay attention to how our stories make us feel.

Here's the thing—no matter how many people (even if they have Ph.D.'s and work at the most prestigious universities) say that the universe is random, indifferent or hostile, it's a story we don't have to accept. The more we tell the story of a friendly universe, the more we look for every scrap of evidence and keep our attention focused on the evidence we find, feeling gratitude for the finding, the more likely we are to live in that friendly universe.

[1]Seligman, M. (1993). "Learned Optimism More Useful than Truth" (interview). Omni magazine.
[2]*Walden*, 1854.

Putting it to Work

What's your answer—is the universe you live in a friendly one?

The point of this exercise is to figure out what your answer has been and what it can be now. You may ask why you need to figure anything out. Can't you just answer the question? Maybe. But the human mind is quite capable of maintaining contradictory ideas (that's what allows us to handle the many paradoxes of our world and the universe) so it's possible that when you give your answer many of your attitudes and much of the way you respond to challenges would suggest a different one.

On the one hand you may say that you believe in the random universe your science education gave you—a world of random mutations, accidental collisions and blind chance, but your expectation of good things, your positive attitude about your ability to handle the challenges of your life, or a sense of being supported and cared for by a higher power, would suggest that there is at least a part of your consciousness that tilts in favor of friendliness. On the other hand, you may determinedly state your belief in the friendly universe you would like to inhabit, while a large part of you is busy ducking and covering to avoid being struck by any of the many tragic events "some unknown force" may launch in your direction at any moment.

Some of your life issues may let you accept a friendly universe while others may give you a sense that there is surely *something* out there working against you, something giving real power to the darkness you see and feel. If you've had a healthy life, a life free of financial hardship, but have never been able to find work you genuinely care about and have had one disastrous relationship after another, you may be okay with the idea of a friendly universe in the first two aspects of your life, but expect struggle in the other two.

Take a look at the Big Stories and Foundational Stories you've found you've been telling in your life so far. How do they fit with the idea of a friendly universe? Remember that a friendly universe has only one force—the force of light, of goodness, of love. There is no opposing force that works against the light, no struggle, no conflict, no polarity or duality. This is not to say there is no struggle in our *experience*, remember—it's to say that it is the *stories* we are telling that create struggle, not the attributes of the universe itself.

Divide a piece of paper into three columns. Write *Friendly, Random* and *Hostile* at the tops of the columns, and then think over the stories you've been telling yourself. Record each of them in the appropriate column. For instance, my old big story about my 50/50 chance of having familial Alzheimer's would go under *Random.* Many of the other stories I used to tell, especially those I told when I was a pessimistic depressive, would have fit under *Hostile.*

I distinctly remember many years ago telling someone who had suggested that living in California might be a good idea, that the moment I set foot off the plane, the state would fall into the ocean. *Hostile,* clearly! One year when my publisher's warehouse had a computer break down just as my book was supposed to be shipped—a breakdown that didn't get fully fixed for six months, I just nodded. Of course that had happened to me, I thought. Given the way my world worked, even my good experiences could get turned into bad ones at any moment. For most of my life the only story I would have been able to record under the *Friendly* heading would have been about my relationship with animals. That was the only part of my life where I didn't feel a need to protect myself from whatever it was that was out to get me.

Once you've had a chance to determine what percentage of the stories you *have been telling* align themselves with the idea of a friendly universe, it's time to address the question of how you can get more of them into the *Friendly* column. The ideal goal would be to get all of them there—to simply and fully accept that you really do live in that universe. (This may take a while, and it's to be hoped that finishing this book will help you do that.) Meantime, you can begin by using your imagination.

What would a friendly universe look like?

This is an opportunity to explore what Einstein might have meant. How would a friendly universe look? What would our place in it be? How would some of the stories you have placed in the other columns work out if they were to be working themselves out in a friendly universe? This is no time to be "realistic," to judge anything by what is unfolding in your experience and the experience of the world around you just here and just now. *Remember that current experience is built from previous story.* Relax, smile, hit your Easy Button and have a good time with this imagining. Just for now allow yourself to build a story where nothing is too good to be true.

Connecting with Nature

Every now and again take a good look at something not made with hands—a mountain, a star, the turn of a stream. There will come to you wisdom and patience and solace and, above all, the assurance that you are not alone in the world.

–Sidney Lovett

How much contact do you have with nature? If you answered *not much* or worse *none at all,* you are missing the single most powerful way to maintain a belief in and an awareness of the friendly universe. It is difficult to explain in rational terms what spending time with nature can do for us. A recent book by Richard Louv, *Last Child in the Woods*[1], suggests that human beings lose something essential when they are divorced from the rest of the life forms on the planet, the natural world of which we are a part.

Make it a priority in your life to get outside, preferably alone, whenever possible in a natural setting, be it woods, shore, mountain, desert, field, creek bank or urban park. Get in touch with the rhythms of the seasons, of the sun and moon and stars. You don't need to *do* anything to reap the benefits of connecting with nature; in fact, it is most useful simply to *be* in the presence of life making its way without the aid or interference of humans and technology. There is a different sense of time, a different quality of existence, a different sound and smell and awareness. An Inuit shaman, Uvavnuk, put it this way: "The sky and the strong wind have moved the spirit inside me till I am carried away trembling with joy."

If the most you can manage is a day or two here and there, a week of vacation, a weekend hike in the woods, that is better than nothing. Julia Cameron, in *The Artist's Way*, advises her readers to take at least an hour a week as an "artist date," to do something or go somewhere they will enjoy alone, as a way of filling their creative well, stirring their artistic juices. My strong advice would be to make a "nature date"—daily, if possible. Take your morning walk in a park, or have lunch under a tree; watch a sunset or the rising of the full moon. Make use of the parks and greenways, the bike trails and office park lakes and lawns, and leave your cell phone in your car.

Bringing the mountain to Mohammed

If there is no way to get yourself physically out into nature, use the power of imagination to bring nature to you. You can collect images that appeal to you, or you can visualize any beautiful natural setting you have visited and can remember. Use your regular meditation practice[2] (or the moments you can take to get away from the hustle and bustle of life) to visualize yourself in the setting of your choice. As always with visualization, do your best to bring all your senses into it. If you are visualizing a stream in a wooded glen, do your best to hear the sound of the water, the birds in the trees around you. Smell the damp earth, the sharp scent of pine. Put your hand into the water and feel the chill and the movement with your fingers. Notice the moss growing on the rocks, the glint of sunlight on the moving water. You can spend a full meditation experiencing any natural setting you choose—the more fully you experience it, the more the effect will replicate the experience of being there. Practicing visualizations of this sort will allow you to make them steadily clearer. And these imaginal[3] journeys have the advantage of allowing you to travel anywhere in the world, to any setting that appeals to you.

It may take a while to notice the effects of your journeys into nature. But if you commit to connecting with the natural world as often and for as long as you can, if you give yourself time to become acquainted with the web of life of which you are an integral strand, the rewards will become clear. As with meditative practice, you may find your breathing changing rhythms, your heart slowing, your senses wakening. The feel of the friendly universe will become second nature. You will begin to know, in a deep and visceral way, that like the trees, the water, the birds and rocks and stars, you belong, and you are not alone.

[1]Louv, Richard. *Last Child in the Woods, Saving Our Chidlren from Nature-Deficit Disorder.* Chapel Hill, NC: Algonquin Books, 2005.
[2]Though I haven't specifically advised a "regular meditation practice" thus far, that too is a way to get in touch with a friendly universe!
[3]"Imaginal" is a word entirely different from "imaginary," which we tend to define as *unreal.* Imaginal means real, though existing only in thought.

10
On Being Selfish

You might as well like yourself. You're going to spend a lot of time together.

–Jerry Lewis

In the last chapter you wrote that maintaining a positive story requires "a steady commitment to our own well-being." Doesn't that make Story Principle really selfish?

Let's deal first with the meaning and context of the term. A *selfish* person is one who, according to the *American Heritage Dictionary*, thinks "chiefly or only" of himself.[1] We humans are a communal species. Thinking chiefly or only of ourselves would seem to deny a critical part of our humanity. We are designed to cooperate with, to help and care for one another. From our earliest history a person who couldn't or wouldn't do that would have been judged a danger to the family, to the tribe. This may be why the teachings of many cultures include warnings against selfishness.

But there's a difference between thinking chiefly and only of ourselves and putting our well-being first, especially since our well-being is so necessarily integrated with the well-being of others. What if what is truly best for the individual is always and forever best for the community? Radical as that seems, it not only fits with the idea of a friendly universe, it is the only way of looking at individual good vs. the good of the whole that *does* fit it. A friendly universe is not friendly to some of its parts and hostile to others. It doesn't support my good while denying yours, or yours while denying mine.

Story Principle is based on the concept that we are all one, that despite its infinite diversity, the deeper reality of the universe is unity. Consider the metaphor of ourselves as cells in the body of humanity. Would you think it selfish for a heart cell to have a steady commitment to its own well-being? The constant pulsing of a heart cell is very different from the task of a nerve cell or a skin cell, but the body needs them all doing what they are impelled to do by their own nature (what gives them well-being), in order to function as an organism made up of trillions of individual cells.

Switching back to the basic story metaphor, we are characters in a Saga being told by Cosmic Consciousness and our own author selves, each of us impelled by our own nature to co-create our storyline as we can. But no matter how closely our storylines are intertwined with those of others, we have the creative power of the author self *only over our own.* As important as it is to consider the well-being of others, given the larger truth that we are all one, to interact most effectively with them we need to put our own well-being first.

What is the announcement we hear every time we fly in a commercial airliner? We are told that if we are traveling with someone who needs assistance and the oxygen masks fall down from above, we are to put on our own before helping the other person. Why? Because if we pass out from lack of oxygen ourselves, we will be no good to anyone else.

Each person, as the hero of his or her storyline, makes choices about how to handle the challenges that are part of that storyline. It may appear that one person's choices can hurt someone else, but in a friendly universe that is only appearance. Remember the concept of the villain helping to create the hero—our storylines intermingle in complex ways. Obstacles can turn into stepping stones and difficult experiences can lead to discovering strengths and developing hidden capacities. Each character is at choice.

Al-Anon, the twelve step program for the families, friends and loved ones of alcoholics, unabashedly calls itself a selfish program, in which each individual is taught to make choices according to his or her own best interests. Often, in trying to imagine what is in the best interest of their alcoholic loved one, or in the best interest of the family, they have been making choices that not only don't solve the problem, but actually help maintain it. This relationship has been termed "codependency," a personal entanglement that makes it impossible to distinguish the best interests of any of the parties, assuring that alcohol remains an ongoing problem and the center of everyone's attention.

Al-Anon suggests that sometimes what looks like the worst possible outcome for the alcoholic (and sometimes for the family as well)—called "hitting bottom"—is often the only route up and out of their bondage to the addiction. Preventing the alcoholic from experiencing the consequences of his or her addiction also prevents learning to cope with it. Keeping the family from the disruption that the alcoholic's hitting bottom may entail may keep all of them stuck in a limited and limiting idea about who they are and can become.

Whatever criticisms of the program one might have, the philosophy at its core— that each individual in any relationship is best able to determine only his or her own best interests — allows all the individuals concerned to make their own choices, to experience the consequences of those choices, and to grow in the process. That doesn't mean that everyone *will* grow. Many people don't want to face the pain and uncertainty that goes with growth and change. But however much we might prefer that those around us be willing to change in ways that would make us happier, we can't make their choices for them.

A steady commitment to our own well-being allows us, whatever situation we find ourselves in and however we got there, to make choices about the only thing we can really fully know—what gives us a sense of alignment with our deepest selves.

The Effects of Being Warned Against Selfishness

Most of us have been warned from childhood against selfishness. The warnings have been well intended, of course—based on the need for human beings to live together and care for one another. The trouble with those warnings, however, is that our natural and necessary ability to love ourselves may well get crippled by them.

The Biblical admonition to love our neighbors as ourselves assumes self-love to be not only a good thing, but the basic reality of who we are. But for far too many of us offering our neighbors genuine love would mean treating them *better* than we treat ourselves.

Women, especially, may have been trained to believe that we are supposed to be "self*less*"—that we're supposed to put the needs of others first, or if not *first*, then at the very least on a par with our own. We're specifically taught that it is selfish not only to think of ourselves *instead* of other people, but also to think of ourselves *before* we think of others. We are given the message that we should take care of others first and ourselves only if time and resources allow.

The message we get on the airplane may make sense to us there, but we are embedded in a strong cultural story that leads us to feel guilty behaving that way in almost any other situation. A friend of mine who retired early and is doing two different volunteer jobs and one for a nonprofit that is barely paid, all of them aimed at helping others, felt guilty for not leaving those jobs and heading to the South to help hurricane victims after Katrina. Another friend, who had chosen to teach in an inner city slum in Chicago, desperately wanted a piano, but felt she shouldn't spend the money—surely it would better be used to help some of her students. Eventually she realized that her spirit *needed* the solace of playing the piano when she got home if she was to have the courage and the energy she needed to help those children day after day.

Okay, but what about parents? Isn't it their job to put the best interests of their children before their own?

One of my favorite bumper stickers (I only saw it once, but it has stayed with me) is "God doesn't have any grandchildren." *Our children are directly supported by the Friendly Universe just as we are.* Of course it's a parent's job to look after the best interests of their children—but there, too, keeping a steady commitment to their own well-being is necessary if they are to do that job well.

When I share this idea with an audience of parents, my experience has been that many loving, kind, responsible parents immediately object—often quite strenuously. It's as if these genuinely caring parents fear that if they are let off the hook for a moment they will become neglectful and their children will be hurt. They seem to suspect that lurking just under the surface there is a selfish monster who might take off for the Bahamas with the paycheck and leave the kids alone at home to starve. (After all, most of us parents have had a moment here or there, when our children are giving us particular trouble, when such a fantasy has occurred to us!) But would they really do such a thing?

Even if they don't think *they* would, they may be afraid that the concept of putting their own well-being first is dangerous because it would encourage other parents to do it. But parents who could genuinely believe their own well-being required abandoning their children are not likely to be parents who are doing their children much of a service by staying home with them.

The Saga we are living in at the level of story is far too complex to allow us to make certain judgments about the outcome of any particular storyline for all the characters involved. We can't know every possible consequence for the children if such parents stayed with them or if they abandoned them. It's possible to imagine ways in which the children of such parents could actually benefit from abandonment. They could be found and removed by authorities or by other relatives. In spite of the horror stories we hear about bad things that happen to children in the child welfare system, there are also real stories of kids who used such a beginning to support a life of success and good works. It's worth remembering that every individual is the hero of his or her own storyline.

The issues involved in parenting are seldom so black and white, of course. What about the father whose feel-good story is that in order to give his family what they need and want, and to be a good provider and worthwhile person, he needs to put his career as an eighty-hour-a-week corporate attorney first, while his children are essentially living without a father in their daily lives? Again, there are so many possible ways the story could play out for each of the characters involved that there's no way to say what all the consequences of that father's story would be, either for himself or his children.

Some fathers live that sort of story for a time, only to discover that it *doesn't* make them feel good over the long haul, so they change it. And some kids whose fathers lived that story the whole time they were growing up decide that being without a father was so painful for them that they would feel good only by finding a way to spend more time with their own kids. Living out a different story, they would have a different relationship with their children and send out different ripples into the Saga.

People have been making rules for how parents should raise their children for untold generations, and whatever rules have been established, they inevitably fail when conditions and cultures change. Conditions and cultures are always changing. So, however comfortable rule-making may feel in the moment, it doesn't work over the long haul or throughout a diverse population! Story Principle says that what works is actually for each individual to do what provides a sense of well-being in the moment.

It's the very best any of us *can* do! When it no longer feels good, we can change it. The bad news is that there is no *right* way to parent, no way that can be guaranteed to have only positive consequences for our children. Remember the story of the father who chose the same gift for both children. Given children's individual perceptions, our parenting choices, no matter what rule or gut instinct we follow to make them, may have a negative effect on our kids we could never have anticipated.

But the good news is the logical corollary of the bad news. There is also no *wrong* way. Even the apparently worst choice may turn out to work well! No right and no wrong, only what feels bad, less bad, good, better, best. All I've said about the relationships between parents and children also applies to relationships with any of those we live with and care about—our spouses and partners, our families, our friends.

Choosing to tell ourselves a story and act within it in a way that feels as good as possible, then changing the story if it begins to feel not good, allows the infinite possibilities in the way those stories might play out to move us inevitably toward the positive rather than the negative. Paying attention to our own well-being is the way growth happens, the way evolution occurs.

Most of us have been given a strong message that feeling good is not a worthy goal (and may, in fact, be distinctly unworthy and likely sinful, selfish and decadent—to say nothing of leading to any number of addictions). It isn't only convent boarding school girls like myself who have been warned about the dangers of making choices determined by a desire to feel good. On the other hand, since Story Principle is based on the idea of Oneness and Goodness (light, love, joy) as the foundational energy of the universe, feeling good is a sure way to recognize that one is in alignment with that universe and one's author self.

Let's consider the possibility that our primary purpose in this life is precisely this, to align ourselves with Who We Really Are, to fully be the person our author self intends. When we feel good, then, we know we are in that moment fulfilling our primary purpose. In this case feeling good, being connected with our author self, is like being plugged into a power source. It can make us feel that we are *more*—bigger, taller, more alive, more capable, intense, powerful, eager, passionate. *Turned on!*

Here's the thing—we're *always* plugged into that power source, and through our Author self to the Ultimate Power Source, the foundational energy of the universe. We can't be unplugged, because that energy is who we are, what we're made of. Even death in the material realm doesn't unplug us from anything other than our bodies.

But we don't always *feel* plugged in, turned on. It's as if there's a circuit breaker that can get tripped so that we no longer feel the power moving through us. We are capable of turning the circuit breaker on to allow the energy to flow again, though most of us don't know that. Even if we suspect we can, most of us don't know precisely how.

As a writer, I was always told that creativity, which is the direct use of universal power, can't be controlled, and I spent most of my life in turn telling others that this was true. Of course, in a way it is true. Creativity can't be *controlled*, leashed, put to work for us like a draft horse under harness—it's a lively, universal force with its own agenda, as it were, its own intentions. It feels as if it comes and goes, dropping in to give us a lovely poem one day, rebelling and disappearing when we are struggling to meet a deadline or conquer a plot problem the next.

"Flow" is the term psychologist Mihaly Csikszentmihalyi[2] coined for the experience of energy moving through us into creative production. It is not exactly like plugging in your computer so it can run and allow you to do your work. It isn't purely utilitarian. Flow is exhilarating. Exciting. Intense. It generates a "buzz." I can think of very little in life that outdoes the high I get when a piece of writing is going so well that words seem to move through my fingertips without connecting in any way with my mind. No effort, no "thinking up" words, no recognizably intellectual process at all. This sort of thing can happen with any creative endeavor. In an interview in *Wired* Csikszentmihalyi described flow as "being completely involved in an activity for its own sake. The ego falls away. Time flies. Every action, movement, and thought follows inevitably from the previous one, like playing jazz. Your whole being is involved, and you're using your skills to the utmost."[3] Note the combination of the ego falling away while the "whole being" is involved. The effect is a sense of getting outside oneself and at the same time becoming more fully oneself, in touch with a deeper, more complete self.

We connect this feeling with creative endeavors because that is when we are most likely to experience it. But it happens spontaneously at other times, too—when we find ourselves in a beautiful natural setting or confronting spectacular piece of architecture, or listening to a magnificent piece of music, for instance. These other spontaneous experiences may be brief and we associate the feeling not with something inside ourselves but with the external scene, object, sensory experience or setting that elicited the feeling. In fact what we are feeling is our own alignment with the universe, activated by the external example of beauty, of harmony. These feel as if they "happen to us," because we may not be aware that we are *designed* to relate to beauty and harmony, to resonate with peace and joy and light.

The Sanskrit greeting *namaste* can be generally translated as "The Divine in me recognizes and greets the Divine in you." It is the beauty, harmony, peace, joy and light within us that is activated when it "recognizes and greets" beauty, harmony, peace, joy and light outside of ourselves. Love is another of those positive attributes of the universe that we resonate with, that we recognize and greet. The feelings we have when we begin to "fall in love" with someone come not solely from hormones and sexual attraction, but from looking at this other person with nearly exclusive attention to whatever in him or her we like and admire. We may not yet have known the person long enough to start noticing or looking for flaws, concentrating on the negative. We are feeling the flow between us of universal, positive energy. We are aligned with the universe.

Any kind of love aligns us with positive energy flow—the love between parent and child, close friendships, the loyalty and commitment that can be part of sibling relationships. Love creates the "heart-warming" qualities of those stories that circulate on the internet of occasions when one person generously helps another; it leads us to admire and respect someone like Mother Theresa. Here again, we tend to think of the positive feelings associated with love as being created by something external. It doesn't occur to most of us that the feelings are caused by something *inside of us* connecting to that external experience, something that is activated whenever we are in touch with the *rest* of who we are, the author self who is not only always there for us, but always in direct contact with and drawing from the foundational energy of the universe.

Like telling ourselves that creativity can't be controlled, we tell ourselves that these powerful positive feelings simply "happen to us" when we encounter certain people, places or objects that bring them about. And we recognize them as brief, fleeting, a kind of frosting on the everyday neutral or negative feelings of normal life.

But once we realize that these feelings come from resonance between the external triggers and the enduring internal reality of *ourselves,* we can turn the process around, focus on the inward reality and, with the positive energy of alignment, draw the external expressions of it into our experience.

A very wise friend wrote that "It is not my responsibility to run the world or save the world. I am a creature of light whose only mission is to enjoy those situations and people who make my eyes light up and to deal gently with those I care for, including myself." She is most certainly not a selfish woman. When her daughter died, she took over the raising of her granddaughter, in addition to her fulltime career. The immediate reality she faced when she made that decision was not a pleasant or comfortable one, but she would tell you that she made it for her own well-being. She was able to tell herself a story of transformation and possibility. It was the choice that felt best in a painful situation and when she'd made it she did her best to head in the direction of joy.

There are people in our lives (our children, our grandchildren, our parents, a dear friend who's going through tough times) whom we choose to take care of because they "make our eyes light up," because we love them, and/or because we feel a deep responsibility to help them, whether we expect the effort required to be comfortable or not. We might not feel terrific making whatever sacrifices we need to make, *but we would feel worse refusing to help.* In caring for them we are very likely to discover that our feelings about the process change—we may go from feeling responsibility as a kind of duty to genuinely enjoying many aspects of our care giving, and we are highly likely to deepen our love for the person in the process. Dealing "gently" with them means that we try to maintain a balance, watching to see what they need or want and then providing whatever part of that we can, while still dealing "gently" with ourselves.

Story Principle doesn't deal with *shoulds.* Heading for what makes us feel better, heading toward the light, one step at a time, leads us inevitably out of darkness. All of us are "creatures of light," but not all of us allow ourselves to know it, to treat ourselves as though it were true. It would serve us well to love ourselves fully, care for ourselves gently, accept our flaws and idiosyncrasies as a part of who our character actually *needs to be* for this story we've entered. Who would enjoy reading a story in which every character was a paragon of human perfection?

The positive *miracle* of self-love is that it connects us to the light within, the light of the Universe itself, and opens our hearts to loving others. Loving others, accepting *their* flaws and idiosyncrasies as part of who their character needs to be, lets them tell their own stories and allows the friendly universe to orchestrate the junctures of our storylines to work as well as possible for all of us.

[1] *The American Heritage Dictionary of the English Language*, Fourth Edition. Houghton Mifflin Company, 2000.
[2] Csikszentmihalyi, Mihaly. *Flow, The Psychology of Optimal Experience.* New York: Harper & Row, 1990.
[3] John Geirland, "Go With The Flow," *Wired* 4, no. 9 (1996): 160–161.

Putting it to Work

1-800-POWER-ON

Where I live our household electricity is provided by Duke Energy (formerly Duke Power). We're surrounded by woods, so when there is a wind or ice storm our power may go out. When this happens, we call Duke's automated number, 1-800-POWER-ON, to report the outage. We get this message from a pleasant female voice: "Your power outage has been reported and crews will be dispatched to restore your service."

When you are feeling "down," when your energy is low or you feel disconnected from the source of your power, you can imagine making that phone call, listening to the reassuring voice that says your outage has been reported and crews will be dispatched (though in this scenario you might substitute the response that "crews *have been dispatched*") to restore your service.

When the electrical power is out in our houses, we moderate our activities. We try to avoid opening the refrigerator, we accept that we won't watch television or work at our computers (at least after the battery gives out). Just so, when you feel yourself disconnected from the power of your deepest self, you can moderate your activities to suit your lack of energy for a time, knowing that it isn't a permanent condition. You can also do whatever you can (as in a power outage we may light candles, find a battery operated radio, fire up the outdoor grill for cooking) to substitute for the full power that is on its way.

Having told yourself the story of making this free call, you have already set in motion the return of full power—the crews are at work. You have only to keep your spirits up in the meantime, generating positive energy in whatever way works for you. You may choose to work with affirmations, to meditate, to listen to uplifting music, read uplifting books, walk the dog, visit an art gallery or museum, climb a mountain, do some yoga—whatever you have found that reliably provides you with a sense of positive energy flow.

The combination of the story that assures you that positive energy will return and the activities that begin to bring it into being, allows you to get back into alignment as soon as possible. Sometimes, like a "real world" power outage, it can return very quickly; sometimes it can take hours or even days. Holding onto the story that the "crews" of the universe are at work can keep you from slipping back into a negative story pattern and prolonging the experience of powerlessness.

When the Circuit Breaker is Tripped

Working with the metaphor of "feeling good" as a sense of being plugged in, a feeling of positive energy running through every cell in your body, an awareness of your personal power, you need to remember that the energy is always available, always there, 24/7. The standard position for a circuit breaker is "on." If the circuit breaker is tripped, flipped to the "off" position, it doesn't mean the energy is no longer available. It only means that you have disconnected yourself from it, from your awareness of it and your ability to put it to use.

Ideally, of course, you would remain conscious of your power in every moment. If you could maintain a story of that constant power, see yourself as a hero in every situation, able to cope "smoothly and easily" with everything and everyone you meet, you could accomplish literally anything you chose to accomplish—you would have only to focus your power in that direction. But life is full of challenges, and your story intersects with the stories of others. You have many old, disempowering storylines stored in your consciousness, some of them long established and not yet or only recently supplanted. You are surrounded by other people who may be living out those old, disempowering storylines, telling them over and over. It can be very difficult to hold on to your power when you're surrounded by people telling a different story about you, about themselves, about the way the world works.

You turn on the television and are bombarded by stories of pain, disaster, inhumanity, accident, cruel fate, looming catastrophe. You visit a family member—a parent or an older sibling perhaps—whose view of you was formed when you really did seem young and powerless, and they mirror that old, familiar image back to you, reactivating it in your consciousness. At work a co-worker who doesn't like you does his best to undercut you at every turn, or a boss refuses to let you take on a challenging task you're feeling drawn to do, because she doesn't think you are up to doing it successfully. You share with a dear friend your idea for a new creative project you're eager to start, and he warns you that it's too expensive, too time-consuming, too difficult, too risky. All these things may trip the circuit breaker and shut down your sense of empowerment.

In addition, since the experiences you are living in the present are a fulfillment of previous storylines, as you attempt to change to a new, more empowering story, the evidence you see around you is likely for a time to support the old stories. Paying too much attention to your current experience and too little to the new story may trip the circuit breaker.

Once you know what disempowers you, you can come up with strategies to maintain your power. You could quit listening to the news that generates fear. Before you go into situations where you've previously felt disempowered, you can remind yourself that what others say of you, what they say to you, the image they reflect back to you is part of *their* story, not yours. It can have an effect on yours only if you allow it, only if you take it in and incorporate it into yours.

You can prepare yourself ahead of time for interactions with others. "Mom is likely to look at my terrific new shoes and instead of telling me how great they look, point out that I'm standing pigeon-toed! I need to remember that she has trouble focusing on the positive with her kids because she never learned to focus on the positive for herself." Having reminded yourself of the other person's storyline, you can go into the interaction not just prepared to maintain your power, but to treat her negative story with compassion. There's a bonus as well. Thanks to the "magic" of Story Principle it is actually possible that having prepared yourself ahead of time to offer the other person love and compassion rather than defensiveness, the sort of negative interaction you originally prepared for might not even occur. You've changed your story and that has changed your experience.

The more we practice protecting our power, seeing ourselves as the hero of a story *we* are writing amidst the Saga of the world around us, the easier it gets. The more we tell ourselves a particular story, the easier it is to suspend the disbelief we may have started with. And the more small, positive stories we tell ourselves moment to moment, with playful, joyful intent, the more we build our overall sense of well-being, and the easier it is to go on shifting all our storylines toward the positive. The more positive small stories we play with, the more we shift our foundational stories. And the more our foundational stories shift, the easier it is to shift the Big stories, so that the Small ones become positive almost automatically. It is an upward spiral that is fully in alignment with the evolutionary thrust of the energy of the universe.

Rebooting

Sometimes you may just have a *really bad day.* You might not even know why. It could be a bad dream with effects that linger, any of the above circuit breaker-tripping experiences, or just an unexplained power outage. Try as you might, you can't come up with a story that allows you to feel better. None of the methods you have tried and found successful to reconnect to your power and get back into alignment works. The very idea of an imaginary phone call to 1-800-POWER-ON seems stupid and delusional.

The first important thing to remember is to avoid beating yourself up. That doesn't help and only makes you feel worse, even more out of alignment with who you really are. The second thing to remember is that not being able to connect with a positive story does not mean that you're inviting immediate catastrophe into your life. One of the good things about Story Principle is that your stories won't manifest in your experience the instant you tell them. You have time to get the story changed! There is that concept of keeping a story active in consciousness 51% of the time in order to bring it into your experience. Even if you slip back to a really grim old story (every so often I have a day of the old depression, so thick and dark I just can't throw it off) there's no need to worry. The more practiced you get with Story Principle the less often this will happen, and when it does you can remind yourself that your life is genuinely better than it was when the old negative story was your primary focus. In fact, that's what makes a bad day feel so very bad—because you've become used to something much better!

On a day like that it's best to relax as much as possible, acknowledge and accept the old, bad story, knowing that it's temporary. *Don't be afraid of the dark. The light will come again.* And, at the end of the day, you can do a reboot.

After a bad car accident, when I'd had a concussion, every single time I fell asleep my recent memory was wiped clean, as if I'd turned off a computer without saving anything. This went on for three or four days. When I'd wake up and discover myself in traction in a hospital I would ask those ridiculously cliché'd questions again: "Where am I? What happened?" My husband finally wrote down all the details he knew so he wouldn't have to keep repeating himself. That was long before personal computers introduced us to the idea of rebooting or clicking "restart." But that's what it was like.

Rebooting a computer can stop a program that isn't working as you want it to and give it a chance to reset itself. You can use sleep like a restart button to give yourself a totally clean slate the next morning. Before you go to sleep tell yourself the story that you will sleep deeply and well and wake refreshed; that any negative story from this bad day will be wiped away and tomorrow you'll be back in alignment with who you really are. Then relax and let the story work.

Set your alarm for a little earlier than usual so that if you don't find yourself reconnected the moment you wake up, you can stay in bed for a while, using that drift time between sleep and full wakeful consciousness, to repeat some affirmations and set a positive story in place for the day. The combination of the intentional reboot with this sort of consciously positive start most often does the job.

11
Control

I am not bound to win, but I am bound to be true. I am not bound to succeed, but I am bound to live up to whatever light I have.

–Abraham Lincoln

It seems that you're saying Story Principle can give us control over our lives.

Story Principle says that we can have *authority* in our lives, which is not exactly the same thing as control over them. Remember that our individual storylines are part of a larger Saga that is playing out around us. In the Cosmology chapter we suggested that Cosmic Imagination creates (in a way we don't understand, just as we don't understand the way our own imagination creates) the Saga of "life, the universe and everything." We are each an aspect of that Cosmic Imagination *but we are not the whole of it.*

If the power of Cosmic Imagination were like electricity, moving from the nonmaterial realm of creation to power the material realm at story level, then our author self would be like a transformer, stepping down the voltage to a level where we can handle it to power our particular storyline.

Consider the Lincoln quote above. "I am bound to be true" means bound to be true to whatever story we are able to tell, to accept, to embody. We are bound to "live up to whatever light" we have, whatever light we are able to bring into our story. That's how it works.

Cosmic Imagination provides the Saga; our author self, through and with us, creates our storyline within it. We have authority only over that, though we *affect* both the storylines of other characters and the Saga itself. Our authority over our own story is, however, complete. The story we and our author self tell within the larger Saga, the story we give our feelings and our attention to, inevitably becomes our life experience. It is what we are bound to be true to and it is bound to be true to us.

Does that mean we can have whatever we want?

It would seem so. But given the essence of what story is, the practical answer is no. Imagine you are reading a book that begins with a character who is completely happy with every aspect of his life. You read the first chapter, in which everything goes *perfectly* for our hero. Whatever he wants he gets. Whatever he plans happens just as he'd planned it and he's as happy with the outcome as he expected to be. Everyone he encounters treats him well. When the chapter ends you begin the next, wondering what or who will throw a wrench into the perfect working of his life, anticipating it with some relish. But in the second chapter things go on as they did in the first. Whatever our hero wants comes to him. All of his plans work. There is no rain when he'd hoped for sun, no sun when he wanted rain. And no surprises. By now you are getting irritated. In the early pages of chapter three the hero's perfect life continues. You flip through the upcoming chapters to see whether anything interesting is *ever* going to happen in this story. Seeing nothing but ongoing perfection, you close the book. What's the point of reading it?

What is missing from this story, of course, is conflict. Suspense. A reason for giving it our time and attention. Without conflict there is no story. In a playwriting class I took in college Joe Stockdale, the professor, insisted that any play we wrote, be it comedy, drama, tragedy or farce, needed to begin with the main character (or characters) "out of adjustment." That was the only way, he told us, to hook the audience and keep them in their seats. If things were going well as the curtain opened, something had to change that in the first scene.

> ***But I'm not an audience member or a reader—I'm the central character of my story. I don't want conflict! I want to have things go exactly the way I want them to. If I can't have that, what's the point of telling me I have the power of creation?***

Have you ever played the children's board game Chutes and Ladders? Most of the adults who have spent much time with that game have done it *solely* to entertain a child under the age of four or five. There is nothing to figure out about winning the game or even playing it. Unless you are playing with loaded dice (in which case you could use some help with your obsession about winning!) you have no control over your progress. You move up a ladder or slide down a chute and one of the players eventually gets to the end and "wins." Not having any control and not needing any strategy is what makes it a splendid game for the child, who has as much chance of winning as the adult. But for the adult all the fun of playing it comes from the interaction with the child—the game itself is utterly mindless and boring!

If life were a game of Chutes and Ladders, and if we were given the "good news" that life is eternal, we would likely end up spending eternity drooling in a corner of the universe somewhere, driven around the bend by boredom and the meaninglessness of it all. It is conflict that makes not just story, but *life* interesting. We need something to figure out. Something to accomplish, preferably over some interesting obstacles. As in a story, we need some suspense. Some surprise. We need an occasional task that we might fail at if the task is to have any value or meaning when we accomplish it.

If you play golf, think about whether you'd play any more if every time you hit a ball it went directly into the hole. How is a golf course constructed? It has woods and water, "roughs" and sand traps, specifically so that the players will face some challenge, some obstacles to getting that ball in that hole! It is those hazards that help them hone their skill. Let me repeat that: *It is those hazards that help them hone their skill.* Some players are so naturally suited for the game and so determined to get good at it that they can become a Paul Hogan, a Babe Didrikson-Zaharius., a Tiger Woods. Others will be perfectly happy as a weekend duffer, just getting out on the course and working on their handicap.

Why do kids (and many adults) get addicted to computer games? Because there are continuing levels of accomplishment that require skill or strategy. When you've managed to work your way through one level, there's another with different problems to deal with. There is the challenge of a problem and the feeling of pride when you meet that challenge.

But those are all "safe" and manageable problems. Life gives us much more difficult and dangerous ones—car accidents, terrorists, wars, lethal diseases.

In a friendly universe our lives are just as safe as the games we have been talking about. Of course our lives have vastly greater emotional impact than a game. Much, much more, including life itself, appears to be at stake. That is why understanding that it's possible to soften the impact by changing our story in any given moment is such a gift.

No matter what the Saga in which we operate provides, we don't have to live a constant life-and-death thriller storyline (though we may do that if we choose, or will if we focus all our attention on a world of constant threats and danger.) But even if we do find ourselves immersed in that sort of story for a time, how we feel about the threats we encounter and how we handle the dangers would still be up to us. The more fully we identify something as evil or terrifying, the more evil and terrifying it will be. You may think you'd like a life free of catastrophes, monsters and demons, but it could be worthwhile to occasionally encounter one of these because of how powerful you will feel when you defeat them or how courageous you realize you have been when you survive.

Many years ago I read Stephen King's *The Shining.* It had just come out and Stephen King was still a fairly new name in fiction. I read the book straight through, sometimes hardly breathing. I told my husband when I finished that I'd felt all the way through as if I'd been tied to the front of an on-rushing freight train that was headed (and I knew it was headed) for a monumental crash. But I couldn't get off. I hated it. But I just couldn't get off. I couldn't shut the book and walk away.

So it can be for someone who is telling a life story, or a piece of a life story, full of intense and painful conflict. As difficult as it is, it is also exciting, suspenseful, dramatic. It demands all their ingenuity and strength. Some soldiers, for instance, say that never again have their lives been as meaningful and important as they felt in the midst of battle. Whatever the horrors of their war experiences, there are some ways in which they miss the intensity.

Recently my husband recommended Stephen King's book, *Cell.* It's an apocalyptic novel and King is a brilliant story teller, so he thought I might pick up some craft tips that could help me in writing *Beyond the Dark,* the apocalyptic third book in my Ark trilogy. I began to read it, and Stephen King began tying me to the front of the runaway freight train again. This time, even though I felt the tug to find out what happens, I closed the book. However powerful King's craft, I no longer choose to give my attention to anything that makes me feel like that. That's the sort of authority we have in our lives.

Story requires conflict, it's true. But there are many different sorts of conflict. When I'd finished *Flight of the Raven,* the story that deals with terrorism, threat and evil, I needed a respite from such intensity. I wrote *Surviving the Applewhites* next. There the conflict is between the two main characters, Jake and E.D., and the wildly creative and chaotic Applewhite family. Jake and E.D. are challenged by that conflict and—as with Elijah in *Flight of the Raven*—uncover new gifts and abilities as they rise to the challenge. But it's a light, funny story because that's the sort of story I chose to write.

The radical and empowering truth about Story Principle is that once we get used to using it, once we let go of the old and pervasive story that our lives are constantly at the mercy of external forces, we can use our authority to choose the kinds of conflict we bring into the story of our lives.

Some metaphysicians say that we come into life to learn lessons. That's one way of looking at it. It's certainly true that we learn in our lives. But to say that the learning is the whole point of living just doesn't work for me. That would be like saying that I write my books in order to learn to write books. No. I write books because of the fun and the difficulty of it, the adventure, the excitement, the feeling of endless possibility and the enormous joy of finally getting them done. Writing books is what I love (and occasionally hate) to do.

Of course I learn a lot with each book, and I enjoy the learning, but only a little of what I learn in one ever applies to the next. Each new book is a whole different set of challenges. For me it's the adventure that motivates me to write. In my days as a depressive, I viewed life as a frightening struggle. And so it was. Now, most of the time, I am able to see it as the same sort of adventure I find in writing. Even when I face the toughest challenges, they somehow feel less challenging than ordinary life as a depressive.

Story Principle tells us that we have the power to create. But it doesn't promise a life without challenges. As we begin to exercise our authority, we can choose not just how to deal with our challenges, but the general shape of them. Remember that the story we and our author self tell within the larger Saga, the story we give our feelings and our attention to, inevitably becomes our life experience. If we would prefer our story to be more like Cinderella's and less like Frodo Baggins', we can tell it from a perspective more in alignment with the friendly universe and less with the struggle between good and evil.

Consider again the quotation at the beginning of this chapter. "I am not bound to win…" If what we are looking for in our lives is alignment with the well-being, the goodness and light of the universe, then the idea of winning doesn't make sense. None of us will ever embody all of the infinite well-being, goodness and light from which the universe arises. "I am not bound to succeed...." That depends on our definition of success. If we choose to see it not as an end result, but as a process, then living up to whatever light we have in this moment, and reaching for more, becomes success. For that we have all the authority we need in our lives.

Putting it to Work

"We must be the change we wish to see..."
–Gandhi

Exercising your authority

Take a moment to focus on the situation in your life that you would most like to change. What you are going to do is to practice exercising your authority over this situation, through a series of steps:

1. Describe the situation.
2. Explain why it is out of alignment with the well-being of the universe.
3. Describe how this situation makes you feel.
4. Explain the changes you would like to see.
5. Examine the changes you've suggested and determine whether they are within your authority to change.
6. If they are not, consider what changes *are* within your authority.
7. Go back over your description of the situation and decide what you need to do to let go of that story.
8. Do that if you can.

9. Revise the story in a way that makes you feel better and returns you to alignment.
10. Watch your new story unfold.

Let's look at an imaginary example. Suppose the situation you want to change concerns a nasty struggle you've been having with a partner in a business venture. Your contract specified that each partner was to contribute his skills to the project and share equally in the profits. But instead, as the project was created you found yourself doing most of the work, while your partner took most of the credit. Because one part of your partner's contribution was to handle the accounting, he has since been in a position to cheat you out of your fair share of the profits, and you feel sure, given the success of the project and the modest returns you've seen, he has done just that. But you've been told that proving it would be a long and costly business that would more than eat up whatever return you might realize. Your interactions have become more and more hostile and unpleasant.

What's out of alignment with the universe is easy to see—in a universe of peace and harmony, nothing about this partnership aligns with those ideals except, possibly, the way the contract was originally written, back at the beginning when the two of you were excited about the project and trusted each other to honor the contract you'd negotiated.

You're feeling misused, abused, cheated and robbed. You're angry and frustrated.

All of that is the story you are currently telling. And it's true. You've got right on your side! The trouble is, having right on your side doesn't change your experience and, if anything, adds fuel to your negative feelings.

What do you want? That's easy too, you might say. You want him to live up to the contract, to give you the credit you're due and to pay you your fair share of the profits. That would put the story back into alignment with the universe, you think, and give you the feeling of well-being you're after.

But your authority over this story doesn't extend to requiring him to change who he is and how he functions in the Saga. Your authority extends only to changing yourself. He is serving as a challenge in your storyline, the antagonist who provides you with an opportunity to discover more of your attributes as hero. You can't force him into alignment with the universe, but you *can* use his presence in your life to expand your own character, to find a way to embody more light.

The new story begins with letting go.

No matter what your new story may be, you can't bring it into your life until you let go of the old one. In this case, letting go of the painful story begins with forgiveness. As difficult as this may be, it's essential. Not forgiving keeps you involved in the struggle, keeps you feeling bad, keeps you out of alignment, and doesn't either punish your partner for what he's done or change how he'll behave in the future. Forgiving him doesn't mean that what he did was okay. What it does is cut the chains that bind you to him and to your painful feelings.

A dog training book I read years ago emphasized the importance of teaching a dog the "Drop it!" command. Your dog might pick up something that would hurt her—a rotting carcass, a splintery bone, something poisonous. Properly trained, she's safe from these dangers because the moment you say "Drop it!" she'll do as she's told and you can remove the offending object. Forgiveness is just like that command. It protects us from harming ourselves by chewing on a dangerous object we inadvertently picked up.

As you're working at forgiving your partner, you might notice the ways in which, during your struggle with him, you mirrored back some of the very attributes you were condemning in him.

No struggle is ever one-sided. This is why Gandhi said we must "*be* the change we wish to see." We cannot bring peace and harmony into our lives by fighting. Having discovered your own contributions to the old, negative story, you then get to forgive yourself as well. You've been doing the best you can and now you are on your way to re-aligning yourself on a higher plane.

Once you've let go of the old story, you can revise the parts of it that made you feel so bad and tell a new one. You can find a way to be grateful for what has already happened. For instance, in this case you could focus on the fact that the project is out there, that you learned a good deal as you worked on it, and you've received some credit and some profit. You know you won't work with this person again, you're clear on the sort of partner you *don't* want in the future, and you can now create an image of the sort of partner you would prefer. Next you can remind yourself that you live in a friendly, abundant universe and set a story in place that your next project will be harmonious, rewarding and profitable.

Then you move to step ten and watching your new story unfold. Don't be surprised if the old partner begins behaving in a less objectionable way.

12
Revision

It's the possibility of having a dream come true that makes life interesting.

–Paul Coelho

You've said that what makes us feel good is in alignment with our deepest selves and with the Universe. But sometimes what feels good now makes us feel bad later.

There's a saying that when the gods wish to punish us they grant our wishes. But that assumes that once our wish has been granted, if we discover that it isn't all that we'd hoped, or that it has a down side we hadn't planned on, or even that we've grown beyond wanting what we once wished for, we're stuck with it. Not so! Remember that as we follow what feels good, *whenever it stops feeling good we can revise our story.*

When I first began to write novels I didn't always recognize an idea that would not only provide the scope for a novel, but was interesting enough to keep me going for the year or more it would take to write it. So somewhere in my files are the first chapters of a couple of novels that I worked on for weeks or even months and then eventually abandoned.

The stories we consciously choose to tell for our lives, particularly when we're new to Story Principle, may be like those abandoned ideas. We get an intriguing idea that we think will take us someplace we'd really like to go. We begin concentrating on that story, telling it as clearly and as steadily as we can, only to discover as it begins to enter our experience that it isn't as useful or as pleasurable as we'd hoped. It takes some time to learn what we really want or even how to think about the subject.

Most of us have been bumping along in our lives, creating by default through the stories that we've absorbed from our past experiences, from the cultural ideas and beliefs we're engulfed in, and from the expectations of other people—like a pin ball bouncing from one obstacle to another. We've gotten used to judging what happens to us by how much we *don't* like it, so that we've built up a substantial list of things we want to avoid. When we try to turn our thinking process around and ask ourselves what we really want, we discover that we aren't sure how to answer. We just haven't had a lot of experience thinking this way.

Once, during a discussion of Story Principle I was having with some highly skeptical teenagers, a young man who was clearly in disagreement with everything I said, spoke up. "It can't be right to just do what we want. If I did what I want, I would blow off my homework, ignore my curfew, hang out with my friends, smoke pot and listen to punk rock."

"And what would happen in your life if you did that?" I asked.

"I'd flunk out of school and get kicked out of the house. I'd have to live on the streets and steal to live, and I'd probably end up in jail!"

"And is any of that what you want?"

"No, duh!"

"So it isn't so much that guiding your story by what you want is a bad idea, as that you have to think a bit about what you really want. You already expect negative consequences from doing those things you mentioned. And even if you couldn't figure it out ahead of time that these things would lead you in a direction you wouldn't like, you'd probably get the message before you actually ended up on the street, if you weren't too tangled up in rebelling against your parents to think it through."

This young man's trouble (aside from being a teenager!) was that he was a lot clearer on what he didn't want—the adults in his life making demands on him, disallowing particular behaviors and controlling his life—than on what he *did*, which was personal freedom and a sense of his own power to make choices. So the first desire he hit on when given the idea he could have what he wanted, was breaking their rules. But, if he had told himself the story that he was *free* and able to make his own best choices, he would have had the latitude to focus on something more meaningful than merely rejecting the wishes of the adults in his life.

If by following what we think will make us feel good we tell ourselves a story in resistance to what we don't want instead of focused on what we do, our story would still be based on the power of what we don't want. We would most likely discover as our story begins to unfold in our lives that we end up with more of exactly what we were resisting. (Note that the boy's first "feel good" story, focused solely on resisting adult coercion, would have led to jail—much more adult coercion that he was currently chafing against.) Resistance is, after all, a powerful focus of attention.

If we discover that a story we've purposely brought into our lives is making us miserable, we can use those feelings to spur us to revise the story. There are other reasons for needing to change our story, too, of course. We and the world around us are always changing, growing, evolving–so what's right for us at one moment may not go on feeling right over time.

A young woman named Kate had had a long term wish for a particular model SUV, but it seemed beyond her reach financially. When she began learning about

Story Principle she decided to tell a story that would bring it into her life. She loved the idea of driving along looking down on other drivers (instead of up at them, as she often did in her old Honda Civic). She was sure she'd be safer in a bigger, heavier vehicle. And she wanted to feel that on her own she was successful enough to drive a big, beautiful, expensive car the way her father always had.

She did lots of exercises to bring the SUV into her life, gathering pictures of it, visiting dealerships and test driving it, imagining how she would feel when she owned one, even coming up with the name she would give it—Petunia. She was careful to focus on the SUV rather than on how she might manage to get it, leaving the question of "how" up to the Universe.

Later that year she got a promotion and an unexpected bonus, and was able to trade her Civic in on the model she wanted, in the classy pearl-white color she loved, with all the bells and whistles that made her feel successful and prosperous. She was thrilled to have it and wildly proud of herself for having been able to make Story Principle work for her.

But only a few months after she drove her treasure home, the price of gasoline sky-rocketed. Her commute to work began to cost her so much that to cover car payments she had to cut back on lunches, movies, and dinners out with her friends. Besides that, global warming had suddenly become a hot topic of conversation and she began to feel uncomfortable driving that big, beautiful gas guzzler. The excitement and pride she had felt gave way to feelings of bondage, fear and guilt. "I would have been better off keeping the Civic," she moaned. "I can't afford to pay for Petunia and I can't sell her. Nobody wants to buy a car like this while gas prices are out of sight!"

For a while she continued to feel bad about being stuck with this car she had so wanted. Then she remembered that feeling bad meant it was time to revise her story. She needed a story that would let her feel better. Looking for a story that feels better is always an individual process. For one person it might feel better to return to the idea of an abundant universe and affirm that since Story Principle had allowed her to get the car in the first place, it could allow her to handle the expense comfortably—and the planet surely must have the same capacity for self-healing as the human body. But for Kate the global warming story was powerful enough to make her feel better going back to a car that was easier on the environment.

The Effect of Feeling Tone

You know how the music in a movie lets you know what's coming? When RJ, our youngest, was little, he would always run away from the television when the music shifted to a minor key. There were scenes in the Wizard of Oz he always missed because the first chord of the accompanying music sent him scurrying out of the room. The feeling tone with which we tell a story is as linked to the way it will unfold as a movie's musical score. If our light, positive story is told through a haze of depression and anxiety, it isn't likely to come to us in a light, positive way. We may get the better paying job we were imagining, for instance, but if our feeling tone was one of fear and desperation while we told the story, something in the new job would reflect that feeling tone. The new boss could end up being a holy terror.

Kate wanted to change her feelings of bondage, fear and guilt as quickly as possible. She didn't want those feelings to affect her new story, which was that she would quickly and easily find a buyer for Petunia and get a decent price, which would allow her to get a car that used much less gas and would also be fun to drive. She'd heard that a playful attitude would help, but she just wasn't feeling all that playful.

So she decided to try a gratitude list. She listed every single thing she could think of that she was grateful for in her life. Once she let go of her focus on how buying Petunia had ruined her life, she found plenty of things for her list. Her feeling tone began to change immediately. She began paying attention to everything she loved about Petunia. She reveled in the bells and whistles, the rich feel and comfort of the leather seat, and the beauty of the pearl-white color in the sun. She kept Petunia washed and assured her that she would soon have a good new home. And she patted herself on the back for finding more and better ways to economize so that she could handle the gas and the payments.

Within a few weeks her story about a new home for Petunia had come about—a neighbor who had been admiring the car from the moment Kate brought her home noticed the For Sale sign she'd put in the window and agreed to meet Kate's price. Before the paperwork on the sale was completed, a cousin who had enlisted in the Army offered to let Kate buy the red Volkswagen Beetle his parents had given him for high school graduation at a very reasonable price.

It's important to remember that with story it doesn't matter if it doesn't turn out the way you'd hoped or planned. Revision is possible any time. With a novel the first draft is seldom the last. If I don't like it when I'm done, I don't keep reading it over and over, beating up on myself for writing it badly. I just read it enough to get a sense of what I could cut out, what I could add, what changes in the basic structure might make the story better. I look for what works and needs to be kept, and try to decide how to hold onto that while creating something new. The underlying threads of Kate's original car story that she needed to keep were her sense that it was okay to have what she wanted, and her ability to bring what she wanted into her life. *What* she wanted was what needed to change.

> ***I've always thought we should accept, even value our feelings. When you say that our feeling tone affects the way our stories play out, it seems as if you're advising us to deny them.***

On the contrary! Denying our feelings would be like taking the battery out of our smoke detector because we don't want to be disturbed by its alarm. Our feelings alert us to whether the stories we are telling are in alignment with the well-being of the Universe. The better we feel, the more in alignment our stories are, and the more well-being we are drawing into our lives. That doesn't mean we will or should feel good all the time; it does mean that our feelings can help us revise the stories we are telling and take us in a direction we'd prefer to go.

It is in the midst of our conflicts and challenges (remember that all stories have conflicts and challenges), that we most need to be aware of our feelings, most need to be able to determine what feels better and what feels worse. By constantly revising our story in the direction of better feeling, we will handle our difficult experiences in the best possible way, step by step from darkness to light.

When my mother was in the late stages of Alzheimer's I was sometimes nearly overwhelmed by negative emotions—grief, frustration, anxiety, resentment, anger—you name it. It wasn't possible to revise my mother's disease or the enormous strain it put on every member of the family. When my father got sick as well and I was the family member living closest to them, I found myself playing a care-giving role I had never expected and certainly never wanted. No matter what story I told myself there was no way the overall situation could change enough to make me feel *good*.

But whenever I felt overwhelmed, I could listen to what I was telling myself in the moment and see if I could make a change that would soften those feelings just a little. Softening the feelings with a new story, however briefly, helped get me, one small step at a time, through one of the major challenges of my life.

One afternoon when Mom and I were taking a walk in her neighborhood, I was suddenly overcome with grief that this woman walking next to me and exclaiming over the same gardens she had exclaimed over on our walk that morning didn't know who I was or where she was. I began to cry, and she patted my arm. "Don't cry," she said. "Whatever it is, it will pass."

In that moment I realized that the story I had been telling—that my mother had "lost" herself, that without her memories there was no one for her to be—could be revised. There was a less painful perspective to take on the current situation.

It was clear that, having forgotten our earlier walk, Mom was fully enjoying the gardens. She was happy exploring this new territory, marveling over the flowers as only someone could who was seeing them for the first time. I remembered the walks I had taken with the kids when they were little, and the fun of sharing their fresh perspective on sights and sounds I had long taken for granted. Why couldn't I see this walk the same way? It was true that Mom didn't know exactly who I was, but she knew I was someone she cared about. And she hadn't lost the impulse to comfort. "Whatever it is, it will pass."

Hearing those words from a larger perspective gave me a different story. This was one moment in Mom's and my long history—a moment that was only tragic if I couldn't accept that our relationship had changed, as everything changes, if I couldn't allow it to be what it was now. This moment, too, would pass, but I could focus on what good I could find in it rather than demanding that it fit into a pattern that would never come again. That new perspective softened my grief. It got me through that walk and that afternoon, and when I could remember it, helped again, over and over, during the weeks and months and years that lay ahead.

Negative feelings alert us to regroup, rethink, change perspective in whatever way we can that allows us to ease those feelings. And as we do that, revising our story to something that feels a little better, our experience improves with our story. What could devastate becomes possible. What could destroy becomes survivable.

Putting it To Work

When a story needs revising

You know when you have a story that needs revising when you find yourself feeling really bad. Story Principle doesn't focus on *why* you feel bad, but on *that* you feel bad. Even if what seems to be the reason for your negative feelings is fully real and even unchangeable—like my mother's Alzheimer's—the story you are telling yourself can change for the better once you figure out just what that story is.

One way to find out is to imagine yourself explaining the *why* to someone, telling them what in your current situation makes your negative feelings inevitable. Write a letter as if to an advice columnist or a therapist, focusing on exactly what's going on and how you feel about it. Fully justify your feelings in this letter.

I told the story about how bad it felt to have a mother with Alzheimer's once a month at the Alzheimer's Support Group meetings I attended. Everyone in the group was dealing with a similar situation, sharing similar feelings, and what I said there never felt like story. My mother had lost herself. It was the horrible, inexorable truth of what happens to people with Alzheimer's. The woman she had been, the things she had been able to do, had vanished.

On that afternoon when I watched her take real joy in seeing the gardens she had seen only hours before and totally forgotten, I realized that much of what made me feel so awful was the story that my mother had lost herself. Suddenly I could see it differently. Yes, it was true that she could no longer do what she had done before, could no longer take joy from the activities she had loved. But joy hadn't left her life. Nor had she lost *all* of who she had been—her impulse to comfort me when I began to cry was a mothering impulse, even though she was not clear in the moment that I was her daughter.

When you have written all you can about the situation that makes you feel bad, probably feeling that you are explaining a truth rather than telling a story, put it away for a day or two. Then go back and read it, thinking of it as story. You may be able to see exactly which aspects of it are creating most of your negative feelings. More importantly, thinking of it as story is likely to show you some things you can change in it to make you feel better, if only just here and just now. Small revisions change your feelings, small changes in your feelings allow you the breathing room to look for more ways to revise.

Revising an old story

How you set out to revise a story depends at least in part on how big a story it is and how long you've been telling it. If it has been a part of your consciousness for a long time, especially if it is pretty well supported by the people and culture around you, the revision may need to be done in modest increments. Leaping too far from the original can make it a little difficult at first to suspend disbelief in the new story. "Yeah, right!" your rational mind may say as you outline a wonderful new story, going on to remind you of all the previous negative experiences that helped you create the one you're trying to change.

Remember Sonja's story about her relationships from p. 107: *I pick duds! Guys who will abandon me just like my father did. No matter how I try to find somebody better, I pick a bad guy every single time.* For her to try to revise that story to *I'm about to meet Prince Charming and live happily ever after* wouldn't be likely to work. Even if that were a reasonable new story, it would be too big a stretch.

Her initial revision could involve verb tense, changing the story from one that stretches out into her whole future, to the story only of her past: *I've picked duds before, guys who reminded me of my father.* She could then add something a bit better about her present and future: *But that's changing now. I know more now about what I really want.*

Imagine this is *your* new story; you could then document that you know more about what you want by making a list of the qualities you would like to have in a partner, and begin giving yourself affirmations about how willing you are to find a different sort of person. These supports could make you feel better and help you keep moving your story along in the direction you want. You could do some gentle day-dreaming about your ideal mate, lover, life-partner, letting yourself get accustomed to the possibility that such a person might exist out there somewhere and might even be looking specifically for someone like you.

In an interview with *Science of Mind* magazine, rock star Melissa Etheridge tells how she found a new love-interest who became her life partner by making just such a list of ideal attributes, while her partner had been doing exactly the same thing. When they found each other, they both recognized the person they had been imagining for themselves.[1]

Keep in mind that revision can be an ongoing process, changing and growing more positive as you feel better and better. And remember the impact of feeling tone. The single most important part of the revision process is to keep yourself feeling as good as possible in the moment. If you can be grateful for at least some of what you are experiencing now, your sense of immediate well-being allows you to keep revising in a more and more positive direction. Sonja wouldn't have to get all the way to Prince Charming and happily ever after to get—one step at a time—from another dud to a loving relationship with just the sort of partner she had chosen for herself.

Be aware of the foundation stones

One of the trickiest aspects of revising a novel is to take notice of the parts of the original story that you are taking for granted—parts you aren't questioning because they've been there from the beginning and seem, as it were, "set in stone." Writing the second draft of my novel, *Listen!,* I was struggling with the fact that the story just didn't feel right. I changed one thing, then another, but nothing solved the problem. From the moment I began the book I had thought that the central character's parents had divorced, and her father had left. One day the phrase "motherless child" popped into my mind and I understood that it was her mother Charley had lost, not her father. That realization set everything right, and the revision found the story that needed telling all along.

You can tear down a tipped and tilting house and put a whole new structure in its place, but if you put the new house up on a cracked, uneven foundation, the new house will end up in no better alignment than the first.

Back to Sonja's story for a moment. Consider what she said about her choices: *Guys who will abandon me just like my father did.* Hiding beneath that description is a tipped or cracked foundation stone that may cause difficulty with a new story if it isn't changed. Whatever is the whole and undoubtedly complex reason that Sonja's father wasn't there for her as a child, feeling abandoned might well have led her to internalize a story that because the father she loved didn't love her back, loving someone leads to rejection and abandonment. It is a story that has gone on playing itself out for her. Until she changes that foundation stone, a new story about finding someone she really wants to have in her life may not work out as she intends.

Like so many Foundational Stories that are formed when we are children, looking at them objectively as adults makes clear how flawed they are. Sonja can find plenty of evidence that loving doesn't necessarily lead to rejection and abandonment. And she could readily see that one specific relationship with one unique individual doesn't have a connection with all future relationships and individuals. Her rational mind will certainly support changing that story.

The only way to keep a negative story playing out in your life is to keep telling it. So look carefully at how you are explaining your negative feelings about the story you wish to revise and see if there may be a deeper story you're so thoroughly taking for granted that you haven't noticed it.

[1]*Science of Mind*, February, 2007. http://www.poetryinlife.com/melissa_main/magazines/science_02-07/science_p1.htm

13
Blaming the Victim

Cancer—you made me what I am today.
–Lance Armstrong

You say that we get to choose the sorts of challenges we face in our lives. What if something really bad happens? Doesn't that make it our fault? Story Principle seems to blame the victim. That doesn't seem fair to me.

Story Principle does say that the stories we live are the result of the focus of our attention—the stories we have consciously or unconsciously told, or accepted from the culture around us. So if something we judge to be really bad happens to us, we must have *some* level of responsibility for allowing it into our lives. None of what we have said so far about the power of positive stories to bring well-being into our lives could work if negative stories did not have power in our lives as well.

The electricity that comes into my house can operate my lights and appliances or burn through a faulty wire and set fire to my house. If the fire happened, the fact that I hadn't fixed the faulty wire would make me in some way responsible—it's my house after all—but if I didn't know there was a faulty wire, I could hardly get it fixed. Being blamed for the fire would indeed be unfair. We may be no more aware of a negative story playing in the background of our lives than of a faulty wire hidden in the walls of our house. So suggesting that when negative experiences result from that story it is "our fault" and that we ought to have kept those experiences from happening, is to demand more of ourselves than is humanly possible.

Before we talk further about the causation of serious challenges in our lives, though, let's consider that term *victim.* It's a term we hear all the time, whenever someone encounters a serious challenge, but it comes from a foundational story of a random or hostile universe. In such a universe we are, of course, at the mercy of external forces. It is powerlessness that creates a victim.

If my house burned down from whatever cause, would I be a victim? Only if I chose to see myself that way. Story Principle says that we are the hero of our story, fully equipped to handle the challenges we face. The house fire would be a big and painful challenge, but I could tell a story about it and about myself that would move me toward well-being, or I could tell a story focused on powerlessness and misery that would move me ever deeper into misery.

We are surrounded by victim stories—people whose lives take difficult turns and the culture therefore labels them victims. Let's take the example of a sexually abused child. The child who experiences abuse has been caught in a powerful cultural belief about the powerlessness of children and the predatory nature of some adults. That belief is the equivalent in the child's life of the faulty wire in the wall of the house. It is not something she is aware of, much less chosen for herself. Her experience undeniably involves pain, fear and trauma. But the effect of that pain, fear and trauma on who she is in the moment and who she becomes as she grows up depends on how she deals with it.

If she and those around her identify her as a victim, that identity is likely to project itself into her future, and she may well create for herself a life story of continued victimization or—in reaction against that victimization—she might become an abuser herself, reliving the victim-predator story in the opposite role. If either of these outcomes happens, her experience will likely be interpreted as further evidence of the devastating impact of predatory adults on powerless children. And so the cultural story continues.

But if she is able to envision herself as the hero of a traumatic story (not necessarily at the time of the abuse, but over time) and determines to find a positive way around her pain, she may use it as a springboard to a life of personal empowerment. Having coped with trauma early on, she may be protected from future difficult experiences by her own sense of personal courage and strength. She has never been a victim, is not one now and will not be one in the future.

Oprah Winfrey is one such person, who has freed herself from the cultural story associated with sexual abuse to become one of the most successful and powerful women in the world. She has used that power to help children and she continues to work toward changing the cultural stories of powerlessness and victimization.

Note, however, that because the predator/victim story is so strong in the collective consciousness, a child who manages to overcome a history of abuse is likely to be seen (like Oprah) not as evidence of the power *every* individual possesses, but as an exception, an anomaly.

The Part the Rest of Us Play in Keeping Victim Stories Alive

To have the life experience of victim we must identify ourselves as one. But the story we are *all* telling can have a powerful effect on how any individual self-identifies. Earlier, the question was asked about whether Story Principle is selfish. Its emphasis on the role each of us plays in creating our own story, along with its reminder that making the best of things and trying always to tell ourselves a story that feels better than any current negative experience, can make Story Principle seem to be the very definition of selfishness—*thinking only of ourselves.* Changing our story and getting a more pleasant life experience as a result would seem to affect us and perhaps a few of those closest to us, but not to have any positive consequences beyond our small circle.

But remember the principle that beneath our diversity we are all one, aspects of one consciousness. As each of us either accepts a negative storyline (whether it exists in our personal experience or not) or chooses to replace it with a more positive one, we have an effect on the cultural story being told at the moment, the story that is being built into the future for all of us. The more the cultural story focuses on the power of evil, of villains, of darkness, and the powerlessness of the individual, the harder it is for any particular individual to get free of that pervasive story. It's like standing near a full orchestra that's playing the 1812 Overture and trying to hum Pachabel's Canon. If we as individuals listen to, accept and retell (to ourselves or others) a negative story about the powerlessness of the individual and the dangers surrounding us, whether from predators, illness, natural disasters or accidents, we add to the cultural nocebo effect that every other person will have to overcome.

On the other hand, each one of us who changes our own story, who focuses on the positive examples that exist to allow us to believably challenge the negatives, makes it easier for the next individual to focus differently. Each of us can choose to look toward light or to give our attention to darkness. So each of us, in deciding what stories to listen to, accept and retell, has power not just in our own lives, but in the lives of others—all others. We are sharing both the Saga of all that is, and the consciousness that creates it. And if that seems a small thing in a world so full of negative stories, remember that any positive story has the positive power of the Universe behind it!

Getting Stuck

Knowing we have authority in our lives gives us the opportunity to use it, but it is a choice we have to make for ourselves. Carolyn Myss uses the term "woundology"[1] for the way many people seek attention by identifying as a victim, retelling the story of the ghastly experiences they have endured to everyone they meet. She suggests that if all of us gave more active and caring attention to each other (more hugs and cuddles) when we are strong, happy and joyful, there would be less need for people to hold onto their pains and difficulties in order to get human warmth and concern from others.

Those who come to believe that they have to be hurt or miserable to get caring attention can have a hard time giving up the story of their pain. The more they tell that story, of course, the more they will go on living it, proving it, creating evidence to cement it firmly in place.

After a car accident that left me hospitalized and in traction with two broken legs and a concussion, my doctor told me that the most dangerous point in my recovery would be the moment when I could begin to walk on my own again. He warned me that after so many months of being cared for, of having people do things for me, it would be hard to give it up and become just another functioning adult, no more special than anyone else. "Some people make it past that point and get on with their lives, and some people don't." I was so eager to get back to my life that I didn't believe him. But eight months later, when my body was ready and able to function independently again, I discovered he was right. I had become psychologically and emotionally dependent on all the extra attention I had been given. I have always been grateful for his warning, which helped me give up the attention in favor of getting back my personal power.

We have probably all encountered people who didn't recover from their own painful story, who replay it over and over, unable to understand why the same sorts of awful things keep happening to them. They are fully convinced that they are innocent victims of the dreadful situations or people they keep encountering. If they could come to a place of enough peace to hear Story Principle and begin to try it, they could take their power back and change their lives. But it isn't our job to tell them that and they wouldn't be likely to hear us anyway. What we *can* do for them is to tell *ourselves* the story that they, too, are safe within the friendly universe, whether they know it yet or not. We can tell ourselves that they have all that they need to change their story, and then treat them that way. Given the power of story, our gentle refusal to buy into their victim identification could do more to help them let go of it than anything we might say.

When Something Really Bad Gets Into Our Positive Story

Chapter 11 said that the authority we have over our own story is complete. And that we can choose whether our story is more like Frodo's or more like Cinderella's. Further, this whole book makes the point that bringing well-being into our lives with Story Principle is a learning experience, something we get better and better at doing with practice. It isn't only that we get better at sorting out what we want and focusing our story on it—it's that we get better at aligning our stories with the light of who we are. As we use new stories to improve how we feel, we feel better about whatever our experiences are, and find it easier to tell yet more positive stories that bring more positive feelings.

But our lives are an almost infinitely complex weaving of story threads, and even the most experienced of us may not be aware of all of them. If, however practiced and confident we are at using Story Principle to create the life we want to live, a massive challenge enters our story or the story of someone we care about—a natural disaster, a serious medical diagnosis, the loss of a beloved spouse or child or friend, a devastating accident—the very first thing we need to do is remove the idea of blame from our consciousness. Blaming ourselves or suggesting to someone else that their extreme pain might have been self-inflicted is not only inhumane, but is a story that makes us or them feel even worse. As we've seen, when we are experiencing negative feelings it's important to do what we can to soften them, not add to them.

This is not a time to try to figure out the cause, which may be impossible anyway. If it was a faulty wire and the house has already burned down, what would be the point? This is a time to stay in the moment, remembering that we can always move in a more positive direction from wherever we are.

Story Principle suggests that no matter how painful an experience may be, there is light in it, good in it, somewhere—because light is the foundation of the universe, regardless of the experiences we encounter at the level of story. That we can't see it in the moment does not mean it isn't there, any more than the storm raging outside proves that the sun has blinked out of existence.

Every massive challenge in one storyline ripples out into every storyline around it, so there are other characters coping with that challenge in whatever way they can. We can help each other to get through it, sharing the best story we can come up with in the moment, or just lovingly being there for each other as each of us copes as we can. It is not anybody's job to sort out how or why this challenge appeared, only to find a story to tell moment by moment that will help us meet it. The more we can hold onto the story of our safety in a universe of light, the better we will be able to cope with the appearance of darkness is our lives and the lives of those we love.

[1]Myss, Caroline. *Why People Don't Heal and How They Can.* Harmony Books, 1997.

Putting it to Work

Perhaps you may kick, moan, scream
in a dignified silence,
but you are so right to do so in any fashion
–Hafiz

That time you felt like a victim

The first part of this exercise is to look back in your life and see if you can remember a time when you felt like a victim. Unless you're extremely lucky, extremely unusual or have a really poor memory, you're likely to find at least one such time.

The one that comes most easily to mind for me is when I was diagnosed with cancer. I did not "kick, moan, scream in a dignified silence." First I got very still and very cold. Then I cried. Then I raged. I distinctly remember watching The Today Show the morning after the diagnosis and being furious at Jane Pauley and Bryant Gumbel, the hosts at that time, for being so obviously healthy. I was terrified and astonished. I'd been blind-sided by the disease. No one in my family had ever had cancer. I had always considered myself unusually healthy. I was only a little over forty! *How could this have happened to me?* There is no question that I felt myself to be a victim.

Having chosen your victim experience, consider what gifts that event or situation may have given you. What happened because of it that would not have happened otherwise? What did you do that you would not have done? What did you learn or achieve as you dealt with it that you might not have learned or achieved without it?

It is not difficult for me to find the gift in my experience with cancer. The very day of the diagnosis a friend whose husband happened to be a radiation oncologist called to chat. She was the first person other than my husband that I told. Immediately she asked if I had a copy of *Getting Well Again,* by O. Carl and Stephanie Simonton. I'd never heard of it. "I'll bring you a copy," she said.

An hour later I began reading the book that set me on a journey that was to change the whole direction of my life. My surgeon was an excellent, but old-fashioned doctor who thought the idea of using visualizations against cancer (as the book suggested) was sheer quackery. He practically crowed with triumph when he told me after the successful operation to remove the tumor, that my twice-a-day half hour meditation sessions had done nothing at all to shrink it. I didn't care. Learning to meditate and using my imagination to work against the cancer cells had taken me out of victim mode and given me a way to do something more than wait for the outcome of the surgery.

Consider how your story allowed that experience into your life.

Now go back, if you can, to the way you were living your life and the stories you were telling yourself before the event you've chosen *happened to you.* See if you can find the aspects of your story that may have brought it about or allowed it to occur.

The friend who had given the Simontons' book to me warned that some people thought it was dangerous because it blamed the victim. She encouraged me to ignore the parts of the book that talked about the possible psychological breeding grounds for cancer and just to use the Simontons' techniques for getting well. But I read all of it and could hardly fail to pay particular attention to what they said about cancer often occurring a couple of years after a period of severe depression or despair. It had been two years since my last major bout of depression—the one that had been the worst of my life, when I had not only contemplated suicide, but had decided how to do it. I had survived that time with the help of an excellent therapist, but it had left me shaken and feeling especially fragile.

As concerned as I was that being a depressive made me more vulnerable to cancer, the curious effect of getting cancer was that I became absolutely determined to survive the illness. It was one thing to take my own life, I thought, but no way was I going to let something else decide when and how I was going to die! The Simontons' book not only gave me meditation and visualization techniques to use in the moment, but reading it made me aware of the connection between my mind and my body. I began to pay attention to the thoughts I was thinking, the stories I was telling myself. And discovered that depression wasn't just *happening to me* either. It was true that I was unusually sensitive to pain, but I also had a habitual pattern of looking at the darkest side of everything in my life.

Getting well from cancer was only the first step. I realized that I needed to find a way to heal the depression that might have contributed to its appearance in my life. Once I had been well for a couple of years, I did get depressed again, but reading that book was the first step in the spiritual journey that eventually resulted in healing the depression and turning my life around. In a way, reading that book was the first step toward writing this one.

Once you have found what you gained from your "victim" experience, you will discover that you, too, have transformed your place in that story from its victim to its hero.

14
Listening Inward

The intuitive mind is a sacred gift and the rational mind is a faithful servant. We have created a society that honors the servant and has forgotten the gift.

–Albert Einstein

Story Principle works because we are not solely the material being we think of as our *selves.* It is the non-material aspects of ourselves, the consciousness of Cosmic Imagination and our author self (that transformer stepping down the power of creation to work within the story) that gives us the ability to create in our lives.

We can think of our character self as our rational mind and our author self as intuition. Rational mind works on the story level—the time/space material dimension. Its perspective is "horizontal," as it were. It processes information that comes to us through our five senses. It works with the logic of the story, memory, and our awareness of linear time that moves from past to present to future. It "remembers" the past and "projects" the future, but it can't know the future except as what appears to be a logical extension of the past and present.

Our intuitive mind functions in the nonmaterial realm outside of space and time. It works with information beyond our five senses and inaccessible to them, information about what is going on in the Saga, including both past and future. It is "on" all the time, operating in all our states of consciousness, including sleep. It knows the intentions, the traits and abilities we brought with us into the world to allow us to make the most of our particular story. And—being nonmaterial and so fully aligned with Cosmic Imagination—it is ready at any time to put the power of creation to work in alignment with our good.

Though our intuitive consciousness, our author self, is aware of information not available to our character self, it's the character self who is in charge of what happens in our lives, who makes our choices. The author can't wield its power unless the character asks it or allows it to. There's good reason for this. The character self is the whole *reason* for our birth into this time-space dimension, the only aspect of us that can have an effect here! Without the character there would be no storyline about us in the greater Saga. Author = Power. Character = Choice about how that power is used.

It seems clear that what would work best for us is for the character self and the author self to work together, coordinating the author self's greater knowledge and awareness with the character self's ability to choose and act in the material world. The question is how do we accomplish that? For many—maybe most—of us the flow of communication seems to be pretty much one way.

We find it difficult to become consciously aware of the author self at all. We move through our lives not actually disconnected from it (we *can't* be disconnected, since it is an aspect of ourselves), but either unaware of it altogether, or unable to create a purposeful connection. It's as if we have an ally with far more knowledge and awareness than we have, a bird's eye view of what's going on, who is always with us, ready, willing and able to help us make the choices that will move us in the direction we wish to go, except that we can't hear its voice. Think of an airline pilot unable to hear transmissions from the control tower. It's as if the radio is turned off.

> ***If I have an author self, I have no way to connect with it. I don't have an intuitive bone in my body.***

That's what I used to say. My article, "From Imagination to Intuition"[1] tells how I discovered my intuition and changed that story. If you have a mind, you have intuition; it's standard equipment. It may be underdeveloped, however, just as a muscle you've never exercised is underdeveloped. It's true that some people are more naturally intuitive than others—from an early age they have a good, clear connection through which information flows easily. But even those of us who aren't aware of this aspect of our standard equipment have experienced its presence and can learn to recognize and use it.

Has there ever been a time when you did something that had an immediate negative result and realized that *something* had warned you not to do it? Maybe you closed the car door on your fingers, or put a cup of coffee on something unstable, so that it almost immediately tipped over and spilled on you. You may have been aware of something warning you not to do it, but you ignored the warning or brushed it aside. Or it felt as if the warning came too late to change the outcome. Afterwards, nursing your smashed fingers or wiping up the spilled coffee, you may find yourself thinking ruefully, wonderingly, "I *knew* that was going to happen!"

That warning was the voice of the author self. If we notice it at all, our rational mind (the character self who is used to being in charge and may think it's all alone so it *has* to be in charge) is likely to discount it or dismiss it. Sometimes what intuition says, gently and unemotionally, makes little logical sense, so that's why our rational mind argues it down. We start out the door on a sunny day and something suggests that we might want to take an umbrella. *Don't be silly,* we think, *there's not a cloud in the sky!* As we stand under the canopy at a restaurant later, waiting for the torrential downpour to stop so we can get to our car, we remember with chagrin that *something* knew the rain was coming.

Intuition is only one of many "voices" in our constantly chattering minds, many of which (the old tapes that are no longer relevant) we want or even need to discount. How can we recognize intuition? The first step is to begin to listen more carefully. Pay attention to all the messages.

We carry with us admonitions and warnings that were given to us as the rules of life when we were children. *Don't go swimming for half an hour after you eat or you'll get cramps and drown; wear your boots—if your feet get wet you'll catch pneumonia; eat that piece of chocolate and you'll get a pimple by tomorrow.*

If a message from inside is one we've heard a hundred times before, if we recognize the voice of whoever gave it to us, or if it carries a load of negative emotion, it isn't likely to be our intuition. Messages from the author self seldom carry a heavy emotional charge. They aren't intended to scare us and they don't involve guilt. They tend to be quiet, gentle, matter-of-fact, provided in calm way, and relevant to the current, specific moment. *Move your hand. That's not a good place to put your coffee. Take an umbrella.*

For many people the messages come in words, a genuine internal voice; for some others they come in impulses that are nevertheless clear. "I started to set the cup down, had an impulse to look for a better place, then put it there anyway," this person might say. Or, "I actually started toward the closet to get the umbrella, then thought I was being silly."

If they are warnings of great danger the messages may come strongly enough that we simply can't ignore them. A friend of mine told me about standing with her mother on the sidewalk in front of their neighbors' yard, admiring their flower garden, when she suddenly heard a loud, intense voice yell, "Get away from the street! Move! Now!" She grabbed her mother's hand and jerked the two of them sideways behind a tree. As they moved a car came speeding down the street, jumped the curb, hurtled across the sidewalk where they had been standing, and came to rest against the neighbors' front porch, shattering the steps. "We almost certainly would have been killed if we hadn't moved," my friend said. There had been no one nearby who could have yelled at her, but she still had difficulty believing that the voice came from inside herself, it had been so *real* and so loud. "Wherever it came from, it scared me so badly I couldn't have ignored it. I didn't think. I just grabbed Mom's hand and pulled her away."

Though the intensity of the voice she heard had frightened her, it wasn't a scary message—it simply gave her the action she needed to take in a way she couldn't ignore.

That friend was not one who had considered herself intuitive. She hadn't noticed messages coming to her about small, daily things, didn't get hunches about big decisions and act on them. But from then on she had no doubt that something real was watching over her.

Seldom is there a need for such intense guidance. Much more common is the message to take an umbrella, to go back for our purse, to make a phone call or send an e-mail, to go check on the kids before we go to bed. We may hear the messages, we may even act on them, without even realizing what we are doing. When we first begin to look for these intuitive messages, we're much more likely to remember the times we *didn't* act on them because when we don't we are likely to get the negative experience that acting on them would have prevented. Those negative examples at least show us the presence of that voice, and just knowing it's there makes us more likely to notice it the next time.

As we begin to actively listen for intuitive messages and to act on them, we are almost certain to discover that more and more of them come. As we pay more attention we can begin to recognize the voice or the feeling we get when it is our author self speaking to us in the same way we learn to recognize the voice and vocal pattern of a person we talk with often on the phone.

Intuition is often called the "still, small voice." In the midst of our busy lives, too much may be going on for that voice to break through and make itself heard. Our need to be constantly thinking of what we need to accomplish next can crowd it out, and our emotional reactions to whatever is happening can overwhelm it. If we can make it a regular practice to get quiet, to sit still, to calm ourselves, our intuition will begin to speak in those quiet times.

It isn't necessary to engage in a formal meditation practice, though that can certainly provide the quiet and calm we're after. Our "nature dates" can do this as well. If we establish that quiet and calm at a regular time each day, intuition will fit itself into that regular practice. I have come to rely on my morning meditative "drift time" between when I wake up and when I get out of bed, to provide access to whatever information my author self might have for me as I begin the day. It's a time when my rational mind isn't yet fully up and running the show. With some regularity there are words in my mind as I wake up—a clear statement that feels as if it comes from someone else—it may have a different vocabulary or a different tone. This is how *All is well, all is well, all manner of things will be well* came to me. Sometimes my author self sends a message in a song. For example, once when I was worried about money "Pennies from Heaven" was playing in my mind when I woke up. I took it as a reminder to let go once again (I have to do this surprisingly regularly) of the old scarcity nightmare.

Any sort of meditation practice can give us calming and centering techniques that provide the quiet we need, but those that go on to insist that we let go of all thoughts may lead us to "let go of" the still small voice as well as the random and wandering thoughts of our rational mind. Since our intention is to turn on the connection with our author self, we need to learn to tell the difference. Here again the intuitive voice will be quiet, positive and unemotional.

If we can't find time for a regular meditation practice we can get calm and quiet at any time throughout the day by withdrawing for a few moments and becoming conscious of our breath. Taking a long, slow breath in and then letting it out slowly, feeling the air moving through our nostrils, and then repeating those long, slow breaths a few times, creates space and silence. We can get off by ourselves even in the busiest day by taking a bathroom break, taking a brief walk outside, breathing mindfully as we go from one place to another or as we handle some task that doesn't take much conscious thought—emptying the dishwasher or waiting for a traffic light to change.

Our author self has a variety of ways to get through to us and some of them will be easier for one person than another. We may get messages in our dreams, for instance. For people who readily remember dreams, this can be an easy method, but it sometimes works even for those who aren't usually aware of dreaming. If a message is really important, the dream that carries it is likely to be both unusually memorable and unusually clear. Dream messages tend to come in symbols that may take practice to learn to interpret, but often the really important messages come in symbols even a two year old can figure out. There are books that claim to provide dream symbol interpretation, but it's wise even if we begin with such a standardized approach to interpret our own symbols because all of us have life experiences that have great meaning for us and give us a set of symbols that are all our own.

Sometimes We Need to Change Our Story in Order to Hear

When our author self communicates by impulse it can be especially challenging to recognize it as a useful message. Most of us have been warned against being impulsive. We are told that children who can't control their impulses have a disorder that needs to be addressed, because acting on impulse is dangerous, undisciplined, uncivilized and threatening to the general welfare. And we've seen the truth of that. A child who impulsively runs out into the street after a ball without checking first to see if a car is coming, can be killed. We are told that our prisons are full of people who have never been able to develop impulse control. And at least some religions warn that our impulses are likely to be sinful or at least come from our animal nature rather than our spiritual nature. So it's no wonder we tend to be wary of our impulses and try to control them.

If we accept that impulses can be intuitive, we can begin to assess them. If we're trying to lose weight and we have an impulse to scarf down an entire bag of Oreos because they taste *so good,* it isn't hard to recognize that impulse as part of the pattern we're trying to break. If we're driving home and have an impulse to turn right instead of going straight, there's no harm in following it. If we then find ourselves passing the cleaner where we need to pick up the pants we'd forgotten would be ready, we can be assured that the impulse was a reminder from our author self. Sometimes we feel the impulse, ignore it, and only after we find the cleaner's receipt do we realize that if we'd turned right we would likely have been reminded.

Suppose we get an impulse to call an old friend we haven't thought of for a long time. We may be barraged by reasons we shouldn't act on it—*it's too much trouble to go find the phone number; he's probably at work and I don't have that number; it'll seem silly to call out of the blue, since I have nothing particular to talk about; maybe he's moved.* But if we follow the impulse and call, we're likely to find there was a reason for it, sometimes an important one. I've gotten in touch that way with my son, who had been wanting to ask me for something he needed but was reluctant to ask for; with a friend who was planning a trip and was delighted to be reminded that she could take a route that would let us have a visit; and with a friend who—that very morning—had been given a cancer diagnosis and hadn't yet had anyone with whom to share her feelings.

A change in vocabulary might help a little. Call an impulse a "hunch" and we may be more open to following it. Whatever we call it, the still, small voice of our author self is offering information that can increase our well-being and maintain our alignment with the positive flow of the Universe.

Sometimes our intuition speaks to us not in words or thoughts or impulses, but with a physical confirmation of something we or someone else has said, alerting us to something that is important, valid, worth following up. A friend of mine gets goose bumps as confirmation. Others get chills down the spine or the back of their neck. I discovered one of my own author self's methods of confirmation when I was in therapy for depression more than twenty years ago, back when I didn't think I had an intuitive bone in my body. Whenever my therapist or I said something that represented an important personal *truth,* my eyes would instantly fill with tears.

It allowed our work together to proceed remarkably quickly. Those sudden tears were like Hansel and Gretel's bread crumbs, making a pathway I could follow out of a dark forest.

As we begin to hear and trust the messages that come unbidden and discover how useful they are, we're likely to want to begin to check in with our author self about decisions we need to make, or to get answers to specific questions. Should we send this note we've just written to a co-worker or wait and talk face to face? Should we speak our mind in a meeting or let it go? Should we apply for that job that sounds interesting? Do we need to follow the advice someone has given us about what to eat or how to exercise?

In the Putting It To Work section that follows some methods of communicating with our author selves are suggested. There are many books, workshops, even college courses one can take to learn how to access intuitive information. What works best for one person may not be useful to another. But all of us can learn.

There is a guidance for each of us, and by lowly listening
we shall hear the right word.
–Ralph Waldo Emerson

We can begin by telling the story that a part of ourselves exists outside the range of our rational minds and five senses, that it is in alignment with Cosmic Imagination and the light of the universe, so represents our highest good, and is always available, willing, eager to share its information. Then all we need to do is listen.

[1]Tolan, S.S. (2006). Imagination to intuition: The journey of a rationalist into realms of magic and spirit. *Advanced Development*, 10, 45-57.

Putting it To Work

Connecting to your intuition—your author self

There are two major ways of connecting. One is to use external tools and the other is to direct your attention to your own consciousness and genuinely "listen inward." For some the second way is second nature, a kind of automatic choice, like the default browser on your computer that comes up whenever you wish to connect to the internet. For those of us who aren't quite sure we have this mysterious awareness called intuition, but are willing to consider the idea that the universe is based on consciousness and unity, it may be easier to begin with the external tools. Think of them as training wheels for recovering rational materialists.

If you're a natural intuitive, you may have difficulty getting past the cultural distrust of intuitive information, the fear you may feel about having access to it, or learning to accept its validity. Many highly intuitive people were told by family members when they were very young to hide their abilities, or they shut those abilities down out of their own discomfort with the information they got or their difficulty with being "weird" or different. Your task is not finding out how to connect with intuition, but how to trust that connection and use it to support your well-being. It begins with telling a new story.

External Tools

Because I'm fond of animals and interact with nature as often as possible, the first external method I used for accessing intuitive information was to take note of the animals I encountered, especially the ones that I would not have expected to see. Someone had given me a copy of Ted Andrews' book, *Animal Speak,*[1] and while it seemed unlikely to me that there could be any connection between my life and the comings and goings of wild animals, I decided to play with the idea that traditional or mythological understandings of the meanings of various animals could provide me with important information. This is a form of "animal divination." I noted to myself when my rational mind rebelled at this game, that ancient people used various kinds of animal divination and found it quite useful. (Remember Julius Caesar's warning about the "Ides of March"—it came from reading an animal's entrails.)

Very quickly I discovered that what I learned from paying attention to the birds and animals that crossed my path really helped me. Mostly what came at first was reassurance that I was on the right track, or a suggestion that I might want to alter the way I was looking at something in my life. The most common experience was to see an animal whose meaning alleviated fear or stress.

For instance, while I was in Arizona for a week, taking a much needed break from a long, tough struggle with my novel, *Welcome to the Ark*, four ravens appeared everywhere I went, including flying back and forth over my car when I was driving. There are four main characters in the novel, one of them represented by a raven. So the fact that I couldn't go anywhere that week without four ravens accompanying me was particularly noticeable. Until the morning I was to head back to the airport, when a solitary raven flew straight across a canyon at me and low over my head, calling all the way, there were never more and never less than four. Among the meanings of raven is "magic" or "miracles."

At the end of that week, when I went home to Ohio where there are no ravens, crows (closely related and almost identical mythologically) took over. For literally months there were crows everywhere I went, as I finished the book smoothly and easily. Had they always been around me and I was just noticing them for the first time, the way you notice the car model you've just bought?

Perhaps. But crows were literally everywhere. I began to think of them as my guardian angels. When I went to speak in a big city crows flew around my hotel rather than pigeons—something I'd never seen before. Wherever I was I'd hear them calling the moment I woke up in the morning.

Then, during the whole month when my mother was dying and I was spending my days at her bedside, the crows suddenly vanished. Search as I might, I saw and heard no crows. Now, instead, I was surrounded by blue jays. The meaning of blue jay? The connection between heaven and earth. The jays were with me every day until Mom died. After that they, too, vanished from my daily experience.

Cards, Divination Tools, and Books

As helpful as the animals were, I decided to purchase a box of Animal Medicine cards. From time to time, when I felt the need of some guidance, I'd pull a card or create a card spread according to the instruction book that came with the cards. Over the years I acquired more sets of cards—Mayan Oracle cards and Osho Zen Tarot cards. I got into the habit of doing card spreads on my birthday and January first, to get a sense of the themes I'd be dealing with in the year ahead.

During the year whenever I felt like doing a card spread, I'd pick the kind of cards that appealed to me that day. Invariably the information I gleaned from the cards was helpful.

There are many collections of cards—tarot cards, daily cards for getting a sense of the day ahead, cards related to certain spiritual or metaphysical traditions—created for varying purposes and representing a large range of interests and aesthetics. If allowing the unity of the Universe to give you some useful information through a random card drawing appeals to you, find a set you like and try them. Trust whatever impulse leads you to a particular set of cards or a particular book—the impulse comes from your author self.

Other divination tools are also available—sets of runes, the I Ching—that come with instruction books for beginners. You can start using them as a game, or you can remind yourself that these methods have been considered useful through many eras and cultures, and may have something especially helpful to offer in a time of satellites and computers when most of us have little contact with the nonrational aspects of consciousness.

In his novel *Illusions,* Richard Bach's messiah figure carries the *Messiah's Handbook*[2] explaining that the way to use it is to ask a question and then open it at random and read the saying found on that page. He points out that it's possible to do this with any book. Scriptures have been used this way for centuries. Nowadays there are even more books and magazines of daily readings of various kinds than there are card sets—so it's not difficult to find materials to use this way, either working through them consecutively or opening them at random. Some of them are specifically spiritual, metaphysical or religious, others are more general and secular. Here, too, allow yourself to make a choice according to what looks or feels right to you.

As you get used to using these tools, you'll find that your intuition not only guides you about which tool to use, but helps you interpret the message, which may be as symbolic as the messages in dreams. But whichever tools you choose, there will be times when they are cumbersome or unavailable when you need them. You may find that you want a tool that's quicker, clearer and more specific.

The quickest, clearest and most specific method I know of is a pendulum. Using a pendulum is a form of "dowsing." You may have heard of people who use forked sticks to dowse for water—it's an old method still in wide use. In many places wells are more often drilled on the advice of a professional dowser than a hydrologist. The idea of dowsing is that we exist in a field of information and a dowsing stick, willow wand, L-rod or pendulum allows the person practiced in their use to access the information wanted.

A pendulum is most often used to answer a specific yes or no question. You hold the pendulum—any small weight on a string or chain about six inches long—so that it can swing freely. Then you ask a question that can be answered with a yes or a no, and allow the movement of the pendulum to provide the answer. There are a number of ways to do this, but all of them involve programming the pendulum to swing a particular way to indicate yes and another to indicate no by deciding how you want it to swing to indicate those answers. Some people like to work with a circular motion—a clockwise circle indicates yes, counter-clockwise no or vice versa. Others prefer a straight line response, a vertical swing for yes, horizontal for no (or vice versa). Some begin with the pendulum still and let it initiate the swing—others start it moving diagonally and let it change direction to provide an answer. Once you've decided what you'd prefer, establish your intention and then ask several questions to which you know the answer, some yes, some no—*Is my name Henrietta? Was I born in London? Did I have bagels for breakfast?* Do your best not to make any movements that would influence the answer. Ask questions until the pendulum swings readily to the correct answer on its own.

It can be argued that even if you don't feel yourself influencing the pendulum, some small, unnoticeable muscle involvement must give you control of the swing. This is almost certainly the case—it isn't that the pendulum is magic or itself has access to the information. The point of using the pendulum is to allow the part of your awareness that has access to information outside of your normal consciousness (your intuition/your author self) to communicate that information. Professional dowsers who work with pendulums go beyond simple yes/no questions by creating diagrams with various answers arranged in a circular pattern around a center point. They hold the pendulum still over the center and allow it to swing to the answer.

My own pendulum came from the Metropolitan Museum of Art and is called Gallileo's Pendulum—it's a plain brass plumb bob which came packed with four answer diagrams. I've never used the diagrams because it's easier just to carry the pendulum in a pocket where it's available any time for a quick yes/no response.

You can make a useful pendulum out of a paperclip or a ring on a string or you can buy one—most metaphysical book stores have a variety, from the simple to the ornate. When I don't have my pendulum I sometimes use a necklace. The pendulum part is easy—creating a clearly worded and fully useful question can be trickier.

But here's what's important in learning to use your intuition, whether through external tools or directly. *The more you trust it and act on its information, the more easily you hear it and the more helpful the information it provides.* As you begin, you may want to ask questions about issues that aren't overwhelmingly important to you, so that you won't be afraid to trust the answer. But the more you trust the answers, the more meaningful you are likely to find your questions becoming.

After you've used a pendulum for a time, you'll discover that just as practicing on a bike with training wheels eventually leads most kids to riding in a balanced way with both training wheels off the ground, practicing with a pendulum allows your intuition to begin to provide the answer before you set the pendulum in motion. You focus on the question, reach for your pendulum and find the answer already in your mind just as certainly as if the pendulum had swung quickly and purposefully to give a response.

Once that happens you will find yourself using the pendulum less and less. But you may still find it useful when a question is so emotionally clouded that you can't get a fix on it, or when it's really important to you but you know you have a strong preference for an answer you aren't sure is the best one. That's when you have to be especially careful to keep yourself from consciously influencing its swing—at such a time I usually hold mine with both hands and my elbows planted firmly on a table top.

Direct Listening

Even if you aren't a natural intuitive, using external tools to begin with gives you more and more direct contact with your intuition. You begin to trust information that you haven't gotten in "normal," rational ways. After seeing how useful the information is, your rational mind is likely to quiet its objections and allow the still small voice to come through more clearly.

Because I'm a writer I am used to having words appear on paper (or on my computer screen) that I didn't consciously "think up." I used to say, *I don't know what I think till I see what I've written.* The act of writing sometimes (not always!) opens a door to subconscious aspects of mind. This first exercise is, therefore, one that works particularly well for me. If you try it a few times and don't get results, don't worry—you just need to try other methods. As with so many things, what works for one person may not work for another. The important thing is to keep trying till you find something that works for you.

Conversation with the author

Begin with a question you would like your intuition to answer for you. This doesn't have to be a yes/no sort of question—the point here is to have a "conversation" about the subject. You can do this by hand (I'd advise beginning this way) or on a computer (probably better only once you've practiced and find handwriting too slow). Write your question and then, consciously changing your handwriting, begin to write an answer. My normal handwriting (created by scores of hours of penmanship practice when I was in grade school) has a "proper" slight slant to the right. When I begin the answer I switch to an upright or slightly left-slanted handwriting. On a computer you can switch fonts. The point is to make a distinction between the source of the writing. There's a commonly used way of getting in touch with your child self—switching to your non-dominant hand for the answer. This is not the purpose here, however, so I don't advise it.

As you begin to write the answer, try to let your hand write it without directing the answer with your conscious mind. It takes some practice. The idea is to relax as much as possible into the "game," and to keep writing. If you stop the flow it is almost always because you're *thinking* about the answer. That means it isn't the author self you're contacting. The author self doesn't need processing time. The answer is immediately available to it.

If you get comfortable with this practice you will often be surprised at the answers you get. The more you do it, the faster the answers will come. That's when you may find it easier to switch to a keyboard so that you can keep up. Or you will get so that you can initiate a conversation just by writing the first question. You'll "hear" the answer and not find it necessary to write it down.

Opening up this channel of communication in such a concrete and direct way may lead you to a more constant interaction. You may discover after a time that you begin to get information even when you haven't specifically asked for it. You have "trained" yourself to be a "natural" intuitive!

Head, Heart and Gut

Another way of connecting with natural channels of communication by listening inward is to think of your consciousness as having three physical centers—the brain, the heart and the gut. All of these physical organs have neurons, all have a "mind" of their own, and all can give you access to different kinds of information.

We can think of the head as symbolic of intellect and rationality. The rational mind is in charge of our survival in the physical realm and so tends to think of us as separate, an individual "I." The heart is symbolic of emotional intelligence and our connectedness with other beings. Its intelligence is "we" centered. The gut is symbolic of our connectedness to our source, our alignment with the universe.

If you have a problem or a decision that you want to process as fully and completely as possible, looking for the answer that supports the "highest good," as opposed to your own immediate self-interest, consider giving all three centers an opportunity to be heard.

Get yourself settled and calm, where you won't be disturbed, and imagine yourself fully inside your head. Describe the situation or ask the question you're wanting to have answered, and let your rational mind give you an answer. Then imagine yourself climbing down a narrow, spiral staircase in your throat and entering your heart. Imagine your heart space anyway you wish to. Repeat the process of presenting the issue and see what the heart's intelligence comes up with. Finally, imagine yourself walking down a broad stairway to a place at the center of yourself, somewhere near your navel. Again, imagine this central location in your body any way you wish. Present the issue again and see what your gut has to tell you. Perhaps all three answers will be the same. Perhaps they won't.

If you are genuinely looking for what will serve your own interests and the interests of everyone and everything connected with you, and if the answers differ, consider choosing in the following way: heart and gut disagree with head, go with heart and gut. If head and heart agree but gut doesn't, go with gut. *The principle here is that heart trumps head and gut trumps heart.*

Practice, practice, practice

Whether you find it easy to access your intuition or difficult, practice will make it easier and more reliable. Use whatever tools work for you. Read as much as possible about strengthening intuition and try whatever feels useful to you. If you give a method a fair chance and it doesn't work, try something else. This is one place where the *only* authority worth accepting is your own. Pay attention to your impulses and notice your immediate response to people, places and things. Make it a regular practice to get quiet, to calm yourself and allow the still, small voice to be heard. It can't be said too often that the more you connect with your intuition, the more powerful its positive effect will be on you and on the Saga around you.

[1]Andrews, Ted. *Animal Speak, the Spiritual and Magical Powers of Creatures Great and Small.* Llewellyn Publications, 1993.
[2]The fictional book now exists in "real life." Bach, Richard. *The Messiah's Handbook, Reminders for the Advanced Soul.* Hampton Roads, 2004.

15
The Death Story

In the depth of the winter, I finally learned that within me
there lay an invincible summer.
–Albert Camus

I've saved this subject for last because it is the most difficult for people to deal with. It is likely that death is *the* negative Foundational Story for all of us beyond a certain age. With the exception of those who have a strong religious conviction that life goes on in some way, death is seen in most of western culture as *The End.* Life is a terminal condition—whatever is born must die, whatever begins must end. So strong is this story, so surrounded by grief and pain, that even many people whose religion gives them an alternative find the alternative to be small comfort when confronted with their own imminent death or the unexpected or traumatic death of a loved one.

Most of our Foundational Stories are established in early childhood, but the one about death tends to get postponed for as long as possible. Children are so focused on life, so eager to experience it, that they seem impervious to the concept of death. Adults, traumatized by the death story themselves, do their best to shield children from it. Should death enter a young child's experience in spite of all efforts to protect him, he may seem relatively unfazed at first. He may ask, "When is Fluffy going to wake up?" or "When is Grandpa coming back?" Such a question seems to adults a clear sign of innocence, of immaturity. Few would see a child's automatic assumption that death is a temporary loss as a healthier and more accurate story than their own.

Eventually, in spite of attempts to protect us from it, the cultural death story intrudes itself into our consciousness where it stays, like a fault line beneath the story of our lives, a fault line that could at any moment bring that story crashing down. However seldom we think about it, we know that whatever lives must die. Our pets are going to die. Our parents are going to die. Our lovers, our friends, perhaps, most dreadfully, even our children. And, of course, we ourselves will die. Every single life that is important to us, *no matter how important*, is headed for death. Either we will eventually lose every person we love, or they will lose us. That's just the way it is.

We may forget the fault line for periods of time, but we are given tremblors almost daily that remind us it's there. The dead opossum on the road, the news of another IED exploding in Iraq, the litany of fires and car accidents, drownings and drive-by shootings on the eleven o'clock news, a friend or family member's grave medical diagnosis.

All of us buy the story that death awaits us and every other living thing. It is obviously true. Scary, dreadful, mysterious, and true.

Regardless of the belief system we may subscribe to about what happens *after* death, our cultural story is that death is the ultimate enemy, just about the worst thing that can happen, something to be avoided at any cost. Politicians, following their own principles or their assumptions about what their constituents want, try to keep frozen embryos alive, to reverse the Oregon law allowing the terminally ill to get medical help committing suicide, to keep family members from unplugging the machines that keep a comatose patient breathing. Protest movements evolve to stop whatever is seen as bringing death to "innocent" people or life forms—drunk driving, abortion, war, pollution, easy access to guns, wearing fur, destruction of the environment on which a species depends.

There is, we are told, a biological survival instinct that marshals the forces of our physical bodies to maintain life whenever it is threatened. We understand that. We have an emotional/psychological survival instinct to match.

One aspect of that emotional survival instinct is a deep fear of death. The story that it is final, an absolute ending, is one of the primary contributors to that fear. If we accept that story, the shadow it casts over our lives is immense. What meaning does our life have if, in the vast expanse of time, it is so brief? What does our learning mean, our growth, our struggle? What of those who die young? What of those who live their whole lives caught in a nonfunctional body or with a nonfunctional mind, or in appalling circumstances of any kind? What of those who die at the hands of another, or in the midst of some project they had worked long and hard to achieve? What could be the point of blinking on and then off again, with no possibility of continuing?

Many of us avoid these questions, not by taking control of our story about death, but by doing our best never to think about it. Some people refuse to make a will or they avoid medical checkups, for fear a doctor will find some sign that death might be closing in. Better not to know, better not to think about it.

Fear of death is, for some of us, particularly focused on losing those we love. Believing that our loved ones end when they die, that they are gone from our experience forever, can make the loss feel unendurable.

In an online community a number of years ago a grieving father began a conversation about the loss of his twenty-three year old son in a drunk driving accident. Reading his posts day after day, week after week, month after month, gave me an education in the devastating power of death beyond anything I had ever encountered. One morning I went to my computer shortly after I woke up, read what the grieving father had written in the wee hours of that night, and had a sudden, excruciating vision of the families behind every accident statistic, every murder report, every sudden and unexpected death in every newspaper in the world, all of them dealing with the intensity of pain that father was describing. It was all I could do to go on with my day. How do people survive this Saga we are involved with, I wondered. How do any of us survive it at all?

As I thought about it, I began to realize that the enormity of the father's pain had at least something to do with his story about death and his story about the meaning of life. If death wipes out not only an individual life, but any possibility of its having had any lasting meaning, then the devastation of loss is compounded exponentially. The father, in losing his only child, felt he had lost not only that child, but any continuity of himself—after *he* died there would be no son to carry on his name, his memory, his genes, his existence. So it was not one ending he was mourning, it was two—his son's and his own. The story of death as ending created emotional devastation that rippled through his experience and into the lives of all those he touched.

Another aspect of our cultural death story is death as failure. The idea of lethal accidents drives us nuts. The liability insurance industry exists because we need to blame someone for failing to protect us from dying. Many people in the medical field look at death as proof that they were not able to do what they were supposed to do, what they were trained to do, what their drugs and their surgeries and their technologies are meant to do—*keep people alive.*

My father's oncologist, who called him Iron Man and genuinely enjoyed being with him as long as he was successfully battling his leukemia, offered him an experimental chemotherapy treatment when a new, more virulent strain of leukemia set in after Dad's initial period of remission. The doctor explained that my father had only a 50-50 chance of surviving the treatment, but he didn't tell him that "success" would only buy him a few more months and those months would be spent unable to leave his hospital bed, instead of at home taking care of my mother, who was in her 18th year of Alzheimer's and couldn't be left alone. The doctor, even though he was an oncologist dealing with one of the most dreaded of diseases, seemed unable to accept that his patients might die. After my father refused the experimental chemo, the doctor never saw him again. He sent instead the woman in his practice who dealt with the *failures.* He needed to devote his time to patients who might succeed in thwarting death a little longer. Though the woman the doctor sent in his place was a good and caring physician who treated him with respect and kindness, Dad remained deeply hurt by what he felt as his original doctor's abandonment.

I was with my father when he died a few weeks later, and felt the extraordinary peace that descended on the two of us shortly after he stopped breathing. It stayed with me for the fifteen minutes I'd been warned to wait before calling a nurse, in case someone might override the Do Not Resuscitate order. That tangible peace changed my story about death once and for all. Perhaps if Dad's oncologist had regularly chosen to be with his patients when they leave their bodies, he would have understood at least that death is not about failure.

For some people fear of death may be made worse because they believe it is *not* an ending—and that they will face punishment for any wrongs they may have committed in their lives. Religions that keep their followers in line by threats of eternal damnation make it hard to treat the idea of eternal life as unmitigated good news.

The problem with our most common death stories is that they are not only intensely negative and frightening, but also intensely resistant to something the stories themselves tell us we cannot avoid, cannot escape. Whatever death is, it is out there in the future for every one of us.

So what does Story Principle offer, then? It isn't as if we can change our story enough to take death out of it!

What we can do is lessen its impact on every aspect of our lives. The first step is to be willing to reconsider our belief that death—as ending—happens at all. Story Principle sees us as larger than our stories, inhabiting the material level of the story as character and the non-material level at which story is created as author. It's perfectly true that as *character* our particular storyline will end. The character will, inevitably, leave the Saga. But character is not all we are. As author we existed before the character entered the story and continue to exist after the character leaves it.

When we dream, our consciousness inhabits our dream, experiencing its story as real, until we wake up. At that point, the dream is over, but we go on. There is sometimes a minute or two of disorientation, as we let go of the dreamscape and re-establish ourselves in our waking reality, but there is no loss of self involved. So it is with our storyline. Our consciousness clothes itself in a body, lives a life in a material world and then frees itself from the body and moves on. Life is not a straight line continuum from a beginning point to an end, but a cycle, ever moving, growing, transforming, evolving.

The question of what happens when the individual life force leaves its material body has been asked by humans forever and answered in myriad ways by different cultures and religions. There are visions of heaven and hell, of joining with a great Oneness, of returning to live life after life. Even when there is agreement on the general outlines of these after death stories, there are almost as many interpretations of how they work out and what they mean as there are people telling them.

In recent times the after death question has been answered by people who have had near death experiences and by mediums who claim to see and speak to the dead. And it has been answered by people like me who have had personal experiences that contradict death as ending. The particular after-death story we tell ourselves is up to us, but allowing an after-death continuation is important if we are to avoid the negative effects on our lives of the death-as-ending story.

For those who believe that consciousness is the foundation of the universe, it would seem clear that death is simply not *possible.* Consciousness must continue through any transformation because it is the foundation of every form, the "stuff" out of which every form is made. If (thanks to visions of judgment) eternal life is what seems most frightening, we can remind ourselves that in a friendly universe the idea of eternal damnation has no place. Whatever our religious authorities may have told us, we get to change *that* intensely negative story if we choose.

Living through the experience of my mother's death gave me a story about the fundamental power, complexity and mystery of consciousness that was a major building block for Story Principle. When my father went to the hospital for the last time I took Mom to a nursing home because our family could not continue to handle both his needs and hers. Over the next four years the Alzheimer's progressed and she declined both physically and mentally. Eventually she fell and broke her hip. By then she had long been unable to follow the thread of a conversation or speak more than an occasional word.

She was on palliative care, and I stayed with her in the nursing home all day every day after her fall, keeping her company and overseeing the way they dealt with her pain. She stopped eating, and everyone expected her to last only a matter of days. She confounded our expectations. Four weeks later, I had to leave her for three days to travel to New England to speak at a conference. When I returned, she had slipped into a coma, but was still hanging on. I went immediately to the nursing home to see her.

Her nurse caught me as I came in and drew me aside to tell me what had happened the night I left. She and an aide had been in the room with my mother when she suddenly sat up in bed. (This was an apparently impossible physical feat, as the broken hip had immobilized her—even the smallest movement had before brought shrieks of agony.) But sitting up was not the only impossible thing she did. She then spoke, loudly, clearly, intelligibly. So boggled were the nurse and the aide that they remembered her exact words: "I've got to go now. Good-bye. Aren't the lights beautiful? Oh—there's Joe!" (My father.) Then Mom lay back down and closed her eyes. She slipped into a coma the next day. Three days after I got home she died.

Some aspect of my mother's consciousness had somehow remained untouched by the destruction of body and brain wrought by twenty-two years of Alzheimer's. In Story Principle, that aspect of consciousness would be her author self, communicating with her caregivers, saying good-by when her character self could not.

Some years later, when I read Rachel Naomi Remen's book, *Kitchen Table Wisdom,*[1] I found the similar story of an Alzheimer's patient who, having had an apparent heart attack, spoke to his teenaged son, telling him not to call 911 and to assure his mother that he loved her and was all right. The man had been unable to speak for fifteen years, his brain virtually destroyed by the plaques and tangles of Alzheimer's. The son became a cardiologist and had never been able to find the answer to the question, "Who spoke?" Western medicine's story provides no answer. These stories and others like them give powerful evidence that consciousness—like life—is indestructible.

An Ancient Metaphor for Transcending Death

Consider the butterfly—one of humanity's most prevalent and enduring symbols of eternal life. If a caterpillar chomping its way through a leaf, should encounter a butterfly, it isn't likely to connect that butterfly with itself in any way. Going by physical appearance they have nothing in common. They don't even eat the same thing! A butterfly is not a caterpillar with wings, a recognizable new model of the old self. It is something entirely *other.*

Let's imagine one caterpillar watching another attach itself to the end of a branch and gradually turn into a chrysalis. Suppose the caterpillar then waits around for a few more hours (a long time in a caterpillar's life span) to see if its companion will "wake up" or "come back." Eventually, seeing no sign of life, it gives up and goes about its leaf-chomping business, mourning (if so it could) the death of its friend.

Eventually, of course, the same thing will happen to our observant caterpillar. Its own chrysalis stage, in which its caterpillar-ness is literally dissolved and reconstituted, must surely feel like death. But the story of universal consciousness suggests that the butterfly, emerging from the chrysalis and spreading its wings for the first time, in some way still "knows" itself and is aware of its own continuity from caterpillar through chrysalis, to butterfly. If it could talk to the other caterpillars munching their way through the greenery, it would no doubt assure them that their future included wings and a taste for nectar! That they couldn't hear it would not change the truth of its message.

Our difficulty accepting continued life after what we call death, given that there is no material form to point to, is not much different from a caterpillar's inability to "see" itself or its kind in the butterfly. It requires that we tell ourselves a story of transformation rather than ending.

> ***But what difference does it make if we believe that our loved ones go on in some other story, when they have left us in this one? The caterpillar you talk about will never see his caterpillar friend again! For those left behind in the story, death* is *an ending!***

Yes. One of the emotional challenges of the stories we live is the grief we feel when someone important—perhaps even critical—to our storyline leaves it and us. They will not again, in material form, walk into our lives and have a conversation with us. They won't be here for the holidays, or to help us through a rough patch. We can feel like a person who has lost a limb—it is gone and we can't do what we did before. It is gone and yet there is the phantom of itself in our consciousness, a phantom that may give us constant pain.

But as we deal with our loss, the story we tell ourselves makes an enormous difference. Months after the car accident in which his twenty-three year old son died, the father in the online community was still—understandably—focused on the pain of his moment-to-moment, day-to-day life without him. Every morning when he woke up his first thought was that Jake[2] was gone. That he would never see him again, talk to him again, hug him again. It was that *never* that made his grief so unendurable.

Someone in the online conversation pointed out to him that if his son had moved to the other side of the world—had gone to Australia to go to school, for instance, or had taken a few years to travel around the world, though he would miss him, he wouldn't greet each morning with such grief and dread. Jake's absence would be as real, but the *story* of it, in any given day, would be very different. It wasn't the *absence of Jake* that wiped the father out every morning, but his story that Jake would never come back.

> ***But if Jake were still alive, he would call or e-mail or write. Having died, Jake could never again communicate with his father.***

Here again, on a day to day basis, the story we tell ourselves either keeps our pain fresh and intense or softens it. Parents whose grown children have established lives far from home, and may not write or call often, aren't likely to mourn on a daily basis the lack of a phone call. Involved in their own storylines, they may look up from time to time and think *the kids haven't called in a while*, but that silence is not an ongoing devastation because they expect it to be broken eventually.

Perhaps they decide to break the silence by calling. If no one answers they leave a message. It's not as satisfying as a conversation, but even a one-way communication gives the parents a sense of connection, and they expect a return call eventually. Meantime, they go about their own lives, their own storylines. In the same way a grieving parent whose story is that death does not really exist may be able to take some comfort in talking to their lost child, even though they don't get an answer, knowing that they will re-connect eventually.

But beyond that, if our story does not accept death as ending, there are ways that we can experience two-way communication with a loved one who has gone that assures us of our ongoing connection. Messages can come to us immediately or days, weeks, months later—whenever we are open enough to receive them.

When my father, a few days before he slipped into his final coma, asked me what I thought about death, I told him that I thought it was probably all right, but I couldn't know for certain. I asked him, if he could, to give me a sign after he died to let me know that he was, in fact, all right. He promised to do it if he could. At the moment of his death, when that peace descended on me, I didn't remember that conversation. Later, I told a friend the story. "It felt like a kind of psychic hug," I said, trying to describe that visceral sense of peace.

"It *was* a hug!" she said. I had told her about my father's promise to reconnect if he could. "He did what you asked him to do!" The moment she pointed it out, it seemed obvious. I had understood, during that fifteen minutes of peace, that death was "all right." I just hadn't seen it as a message from my father. If I had already had a firm story in place that death was only transformation, I might not have merely *described* the feeling as a psychic hug, I might have recognized my father's presence and actually *felt* it as that.

So important is our story to our experience that if I had fully believed the story that death is the end, I might not have felt that peace at all and would have missed the opportunity to look at death in a new way. If I had missed it, of course, the story of death as ending would have been reinforced.

There are many other ways that those whose stories allow continued connection to their loved ones after death can actually experience that connection. About a week after my mother died I had a dream—the kind of clear, utterly memorable, *real*-feeling dream that people often report after someone they love has died. I drove up the driveway to a bed and breakfast and my father—

looking tanned and fit and very like he looked when I was a child—came out on the porch to greet me. "Your mother's inside," he told me, "unpacking. Would you like to come see her?"

"Sure," I said, and went inside with him. In a lovely bedroom my mother, in her slip, was taking clothes out of a suitcase and putting them into an antique dresser. She, too, looked the way she had when I was a child. I greeted her, and she turned and smiled at me. Then she went back to unpacking. It was clear she was glad to see me, but also clear that she had other things on her mind and didn't want to stop what she was doing to talk to me. There was no sign of Alzheimer's. Both of my parents looked fit and healthy and happy.

When I woke I had no doubt that the dream had been a message—from both of my parents—that they had gone on to a new storyline, together.

Sometimes, because adults believe so firmly that there can be no connection between the living and the dead, the messages from departed loved ones come through children, who are more open to receiving them. One family I know was gathering for breakfast when the youngest daughter, aged four, came into the kitchen. "Grandpa just came into my room," she announced. Her mother nodded absently, assuming that her highly imaginative daughter was telling another of her stories. Grandpa lived more than a thousand miles away.

"He told me to tell you that he loves you and that you shouldn't worry about him. He's okay. He told me to come down *right now* and tell you that."

A few minutes later the phone rang. The call was from the child's weeping grandmother, saying that her husband of fifty-two years had had a sudden heart attack and died.

For some people the messages come in symbols that seem particularly connected to the person who has died. There are stories of dragonflies appearing where no one had seen dragonflies before, a symbol of the presence of a woman who had filled her home with dragonfly images. Or coins turning up in strange places—in a refrigerator, on the table at a family dinner, in a cupboard, in the dog's bed—symbols of a coin-collecting father who had died.

A woman whose live-in companion had recently committed suicide was having dinner with a group of their old friends. They were all talking about how much they missed him, how their group had changed since his loss, doing their best to help her with her grief. The waitress took their drink orders. When she brought the drinks, instead of the cosmopolitan the woman had ordered, the waitress put in front of her a cranberry juice with vodka—the drink her companion had always ordered. No one at the table had ordered it. When the waitress tried to find who the drink belonged to, no one in the restaurant had ordered it. Everyone at the table agreed that drink had been a message from their friend.

Story or reality? To them it didn't matter. They had felt the presence of the friend whose loss they had been mourning, and it helped them all. As difficult as it is to endure the loss of those we love, a different story about death can help us through it.

At the Center for Grief and Traumatic Loss in Libertyville, Illinois therapists say they are able to facilitate after death communications by using sensory desensitization and reprocessing to induce an altered state of consciousness with rapid eye movements. In Ervin Laszlo's book, *Science and the Reenchantment of the Cosmos*[3], he cites this work and asks whether these communications, which have great therapeutic value for the grieving clients, are *real.* He then goes on to quote a letter from Dr. Allan Botkin, the head of the center, saying that these communications have convinced him of the survival of individual consciousness after death. Dr. Laszlo finishes his exploration of the question of immortality by saying that science does not know how this could be, but that it "seems to be the case."

Here again, Story Principle does not, like science, deal with the question of *real* or *unreal.* It sees all experience as story, so the question to be asked is whether the story we tell provides for our well-being or not. Over and over in the literature on near death experiences and after death communication we see that the story of the immortality of the individual self makes life easier, alleviates pain and grief, and helps us handle the inevitability of the end of our own or anyone else's storyline.

[1]Remen, Rachel Naomi. *Kitchen Table Wisdom, Stories that Heal.* Riverhead Books, 2006, pp. 300-301.
[2]A pseudonym.
[3]pp. 73-77.

Putting it to Work

Exploring your death story

Keeping in mind that Story Principle does not deal with real or unreal, but with the effect upon us of the story we tell, this exercise encourages you to explore the death story you have been living with and evaluate its impact on your personal well-being. You may do the exercise in writing, or you may simply look back at your life experience and process it mentally. If you aren't ready to deal with the subject at all, you might want to consider what this tells you about the death story that is running in the background of your consciousness.

Your earliest memory

What was the first encounter with the death of a person or animal that you can remember? You aren't necessarily looking for the first death that touched your life, but the first that you can clearly remember. Think back to that experience and relive as much of it as possible. Consider the following questions, answering any of them you can.

How did you feel about the death? If you were a child, what did the adults in your world tell you about it? Did they encourage you to process it, or did they surround it with silence? Were you given an opportunity to participate in a ritual around the death? If it was an animal, was there a burial? If it was a person, were you included in a funeral or memorial service? How did these rituals, if you experienced them, affect your response to the death? Were you given a religious explanation or interpretation of death? Were you told that there would be a time in the future when you and the person or animal would meet again, or were you told that the person or animal was gone forever? Did you feel that your questions were answered? Were you comforted by whatever answers you were given?

Other encounters with death

Now consider other deaths that have touched your life since. Were you helped by the death story you created or were given originally? Have you changed your story since? If so, what led you to make those changes? Do you think the changes make you more or less comfortable with the prospect of death for yourself or others?

Do you believe it is possible to reconnect with someone who has died? Have you had such an experience, or do you know anyone who has? If you've had that experience, how has it affected your death story?

If you have experienced the death of someone very close to you, are you still grieving? Does your grief affect your daily life or has it moved into the background to some extent? Do you expect it continue to be a presence for the rest of your life? Do you think it is possible that a change in your story about death could soften or shorten the grief process?

Your story today

What is the story you tell yourself about death now, and how does it make you feel? If you see death as an ending, does that story distress you or give you comfort? If you accept a story of eternal life, what does that mean to you? If you're uncertain about what death is about, and what, if anything, happens afterwards, are you comfortable with the uncertainty? How would you describe your emotional response to the prospect of your own death?

Considering changes

If, after exploring your attitudes and feelings about death, you would like to create a more positive story, you may do exactly that. Just as you earlier imagined what a friendly universe might mean to you, see if you can let go of the most negative aspects of your story and replace them with something that makes you feel better.

Remember, as always, that you are totally free to tell this story any way you like. You aren't struggling to figure out what is real—you're allowing yourself to tell the story that will soften your attitude about this inevitable human experience.

Though you may not have recorded the other parts of this exercise, it might be a good idea to write down the way you would like death to be, what you would like it to mean. You may someday wish to revise it in an even more positive direction, but it is useful—given the power of the death story in the world around us—to have the most positive story you can imagine today to come back to and reread. It can't be said too many times that the story we tell is the story that becomes our experience.

16
Death as Choice?

To take the first step is to close one's eyes and dive into unknown waters for a moment, knowing nothing, risking all, but then discover the waters are friendly.

–Rabbi Sam Apisdorf

Since Story Principle says the story we tell creates our experience, does that mean we choose our own death?

This is an emotionally charged question. Before attempting an answer, let's frame the context to deal with some of that emotional charge. In our culture a chosen death is known as *suicide*, a word and an act fraught with trauma. A person who chooses to end his life is most often seen as running from difficulty, walking out on his possibilities, turning his back on his obligations to his own life and the lives of anyone connected with him, and betraying those who love him. Suicide is considered tragic, wrong-headed, even evil. In a Judeo-Christian context, suicide is seen as an affront to the God who gave the person life.

Even a choice to die by a swift and pain-free method with the help of a doctor rather than enduring a long, agonizing death from illness is called physician-assisted *suicide.* So strong is our cultural negativity to suicide that the federal government has attempted to override the results of a democratic vote in the one state where this practice has been legalized. So first we must disentangle the question of death as choice from the heavy, negative connotations of suicide that upset us so.

There is one version of a chosen death that is already free of those connotations—the heroic concept of giving up one's life for the life of another. When a soldier throws himself on a grenade, or a drowning mother manages to put her son on the inner tube that could have saved her, their deaths, though chosen, are not labeled suicide. As painful as their loss may be to those who cared for them, there is an offsetting story about their noble sacrifice—even if the person they sacrificed their life to save was a stranger—that helps to soften the survivors' grief. If we keep in mind that there is this example of death by choice that is accepted, even glorified, we may be able to free ourselves to look at the question with a less negative perspective.

Now let's put the question into the context of the death story suggested in the previous chapter. If death is not an end to life but a transformation, a change, a moving on to our next life experience, more in the nature of changing schools or neighborhoods or careers than giving up our very existence, then suggesting that we might have some kind of choice about it doesn't have to be quite so troubling. In fact, it begins to make considerable sense.

We understand and accept, however grudgingly, that every person born into this life experience will also leave it. So the idea of choice does not include whether to die or not, but only when and in what circumstances.

When my father chose not to try the experimental chemo that might have prolonged his life by a few weeks or even months, he was exercising choice over both timing and circumstance. Knowing that his death was inevitable and imminent, he was not willing to attempt to stretch out the time he would be incapacitated and dependent. All people who sign a Do Not Resuscitate order or a living will are making conscious choices about the timing and circumstances of their deaths. They may make their choices to maintain some quality of life; they may make them to protect their families from the financial burden of their final care, but we accept their right to choose for their own reasons. Except in the case of underage children or adults who have been judged incompetent, most of us believe that individuals should be free to refuse medical care, even medical care that would or could save their lives. A condemned prisoner may be allowed to choose the death penalty over life in prison.

These are all conscious choices about the timing and circumstances of death; our willingness to accept them is based on the idea that the person making the choice is mentally competent to do so. We allow others the right to choose in these situations because we wish to have that right ourselves. We want as much power in the face of death as we can possibly have. Story Principle can give us more than most of us have ever imagined.

What Story Principle says about death is the same as what it says about every other aspect of our lives—our story, either purposefully told or simply allowed, consciously or unconsciously, determines our experience. Positive stories fueled by positive emotions—joy, love, gratitude, peace—create positive experiences, stories fueled by negative emotions create negative experiences. Consciousness creates reality whether we are going to like that reality when it arrives or not. As with other issues, we may be choosing without knowing we are choosing. Hostility, anger, fear, depression, hatred, jealousy, resentment, grief, despair—stories growing out of these emotions or activated by them can lead not only to actions or behavior with harmful or even deadly consequences, but to physical illness that could be life-threatening.

Remember that we are both character and author in our story, and that our author self is in alignment with our good at all times, though our character self, caught in the narrower perspective of story, may not be. Our character self may smoke, drive at excessive speed, sky dive, set out to climb Mt. Everest, pick up strangers on the highway. This is not to say that any of these activities is a choice to die—many people do these things and live long productive lives. What determines the outcome of risky behavior is not the behavior alone, but the story we tell about it and the positive or negative emotions involved. A joyful and care free sky diver may well be living a less risky life than someone who is terrified that death might lurk around any corner and so devotes most of his attention to protecting himself. In any case it is the character self that is telling the story and doing the choosing, with or without consultation with the author self.

The life story we are telling may move us toward its ending in ways we would not consciously choose, but we can change that at any time by changing our story.

What about murder? What about a plane crash where hundreds of people die at once? What about tsunamis or hurricanes or earthquakes? Did all those people choose those deaths?

It is likely a different story for every one of them, no matter how many people are involved. Many who die in mass events may simply have unconsciously allowed it. We are all embedded in a Saga in which death occurs in many different ways, so in the background for all of us there is the possibility of murder, accident or natural disaster. Not creating a story that *disallows* such circumstances for ourselves would not necessarily bring them into our experience, but it *would* keep all of them possible.

There are people who may have been telling a story that would have led them to their deaths but who changed it somehow so that they survived instead. Someone who was planning to take a particular plane or train decides for some reason not to go and then there is a crash where many or all of the passengers die. Studies of train crashes have shown that there are fewer passengers on the trains that crash than normally ride that train on that particular route. On 9/11 so many people who would usually have been at work when the planes crashed into the Trade Towers weren't there that a conspiracy theory arose to account for it. And there were people who, though they had chosen to spend their holiday where the tsunamis of 2005 happened, had decided to do something that day that took them safely away from the shore. These are the ones who heard the whisper of intuition and heeded it, whether or not they recognized it as such.

Some of those who die in such events may have had a similar warning but dismissed it or couldn't hear. Remember the still, small voice that says, "don't set the coffee cup there" but we do it anyway. As with other aspects of our lives, our author self, having attempted to communicate its greater knowing, does not over-ride our choices. If this seems frightening it is because we are caught in the story of death as ending.

The author self understands, from its place of alignment with the fundamental goodness of the Universe, that no matter what the effect of our choice may be on our current storyline, *including ending it*, we are ultimately safe. It is impossible for us to end our existence.

There are almost certainly others who die in a mass event who have made a decision in alignment with their author selves to allow their story to end at this time in this way. Their reasons will be as diverse as their personalities, their temperaments, their individual storylines. We can neither know nor judge from the outside. In some cases whole families die together, and as upsetting as many of us find that, there are worse circumstances than going into the next life experience with those we love, leaving none of them behind to grieve alone.

But what about those who do get left behind? Spouses? Children?

This question may seem to be a question about death, but it actually returns to earlier questions about facing the inevitable challenges of our life stories. The death of someone we love or depend on is a challenge that may call out every bit of strength our character self can muster. And, of course, it challenges us to decide what story we will tell ourselves about death.

Here's an example of apparently devastating circumstances that led to greater possibilities instead. It's the early life of Alexander Hamilton, as reported in Garrison Keillor's *Writer's Almanac* on January 10, 2007: "He grew up on the tiny island of Nevis, where his father abandoned the family and his mother died when he was just a boy. But he was taken in by a local merchant, who gave him a job at a general store. He turned out to be quite good at accounting, so when he was 13, his boss took a trip to Europe and left young Alexander in charge of the store. He started writing on the side, and an article about a recent hurricane so impressed the adults around him that they all pitched in to pay for his passage to New York, where he could attend school."

We need to remember that our sense of tragedy about the events of someone else's life is just *our* story about those events. Each person must create his own. The challenge of abandonment and death are the very events that moved Hamilton off his tiny island and into realms of accomplishment far beyond any normal expectations from such a beginning.

What about the death of a baby or a child? Are they telling that story?

The author self of a baby is not a baby. The author self of a child is not a child. That's a tricky concept, but outside the realm of the story our consciousness is ageless and timeless. It is simply who we are. So, while a baby doesn't seem to be in a position to tell itself a story of any kind, its nonphysical self knows its intention for entering the Saga and is aligned with its good as much in infancy as at any other time of life. A baby is not likely to be fully aware of itself as separate in any way from its nonphysical consciousness. So it is more likely than any of us to experience life in full alignment with that consciousness.

Additionally, however, the baby's storyline is enmeshed in the storylines of its parents and all the other people in its brief life experience—so their storylines, including their fears and unconscious negativities—may play out in the baby's experience as well as in their own. Infants and children seem able to slip more easily out of physical existence than adults, perhaps because they are closer to the realm of light beyond story and so are less fearful of going back there, perhaps because they have less investment in whatever story they have begun to live. And here again, if death is not an ending, an early death is not an early *end.*

A child may in fact choose, in cooperation with her author self, a storyline that includes only the experiences of childhood. Though most of us may think a full and satisfying storyline requires the whole trajectory of human development, from infancy through old age, that is not the only story worth playing out. Any life story, with any ending, may seem unfair, unacceptable, cruel or wasteful to those observing it from outside. It is not necessarily any of those things to its central character.

Can we consciously choose the story we tell about our death and bring it into our experience as we've told it?

Yes. We can. The more difficult question for each of us is whether we can *accept* that we can. My mother-in-law, living on her own at age 93, began to be concerned about how her storyline would end. She watched her husband of nearly 68 years dwindle, loose his ability to get around, to hear, to see, to care for himself. When he was admitted to the hospital for the last time, she was told that he was not going to get better. She stayed with him, watching and fretting as they woke him to take blood, put needles into him, gave him pills, performed tests that would do nothing to save his life or even delay his death. All those things were simply part of the hospital's method of patient care.

She didn't want that to happen to her. Her living will, she knew, didn't disallow taking her to a hospital if she fell and broke a bone or had a stroke or a cardiac problem. So she was afraid of all of those possibilities. What she wanted to know was that she was not powerless over the standard procedures of modern medicine, that she could tell the story of her death any way she chose.

So I asked her what that way might be, and she told me that she wanted to slip away in her sleep. I told her I thought that if she wanted to consciously tell that story and could manage to override the memory of all the people in her life who ended their lives in hospitals or nursing homes, she could do it. I explained that her task was to suspend disbelief in her positive story, focus on the good and enjoyable things that remained in her life, and trust that the friendly universe would bring her the experience she had chosen.

While I was writing this book she found herself in the hospital suffering from congestive heart failure. Her doctors said she could no longer live alone, and so my husband and I invited her to move in with us. She was not expected to live out the month (this was June, 2007). Two weeks after she moved in, a medical evaluation showed that she was not eligible for hospice care. Being surrounded by family and loving care, her health had already improved dramatically. She lived with us, visiting regularly with her daughter for a few days or weeks at a time, for 19 months. In January, 2009, at the age of 95, she died peacefully in her sleep. The story she had chosen to tell was the story she experienced.

What if we all knew that we could choose the way we leave our story? It isn't likely we'd all choose the same circumstances or the same timing, partly because we are individually very different from each other, and partly because the stories we have created are very different. In his novel *Illusions* Richard Bach's "reluctant messiah" says of death: "Dying is like diving into a deep lake on a hot day. There's the shock of that sharp cold change..."[1]

When we go swimming some of us like to dive right in, enjoying the exhilaration of that shock and the sudden transition from hot air to cold water. Others of us creep into the water a little at a time, getting used to the chill one body part after another until we are finally fully submerged. Which is the better way? Whichever way suits us.

In just such a way some people would prefer a sudden exit, in the midst of their lives when things are going well; others aren't willing to let go of this life story until the last of its good parts are sucked dry and what is left is no longer worth living. Some would prefer to leave at a high point of action, intensity and drama, others would rather complete the action part of their lives, rest a while, basking in their memories, and then slip away gently with a sense of completion. Some would want their loved ones around them, sharing the last moments, others would hate having to say good-bye. However any of those choices might look to others, they are choices people might consciously make if they knew they had the power to do it.

Given the emotional weight of the subject of death, the power to choose the timing and circumstances may be for some of us Story Principle's greatest gift.

[1]Bach, Richard. *Illusions, Adventures of a Reluctant Messiah.* Random House, 1989.

Putting it to Work

I'm not afraid of death, I'm afraid of dying.
–Woody Allen

Daring to imagine your own death

If you didn't wish to engage in the exercises about death after the last chapter, you may have a similar problem with this one. Or you may find that doing this one makes it easier to directly confront the story you have been telling yourself about death.

There's no wrong way to do this—if you aren't up for dealing with the subject of death at this time, you can return to the exercises later, or skip them entirely. It is possible, however, that this exercise is the most liberating one in the book!

You can do it, as usual, in whatever way suits you best, but you'll want to be able to revisit it. You can write it in your journal or on your computer, you can draw or paint a picture or look for images in magazines or on the internet, or you can make an audio recording—anything that will give you a way to read, look at or listen to it again. It's a good idea to date it so that you will have a record of when you created it.

Begin by affirming, whatever your death story may have been before, that you can't actually die.

Death as we know it at the level of story is not an ending, but a transformation, no more threatening to your beingness than waking from a dream. Remembering that you are both the author and the character in your particular story line, you are about to invoke your ability to choose how your character self will complete this story.

What you're going to do is create the story of your death *as you would like it to be.* As always, when creating a story, you are free to create it any way you want. It can be dramatic and heroic or peaceful and quiet; sudden and unexpected or gently and lovingly processed. If you'd like to be surrounded by loved ones, tell it that way. Give the story whatever setting you wish to create. You can tell it in first person or third, letting yourself be as close to the telling or as distant from it as is comfortable for you.

Remember that you're telling this story your way and you may give yourself as much or as little lead-up to the moment of death as you wish, and follow yourself as far as you'd like, out of this story and into whatever you want to come next. You can draw on your own imagination, on mythology, dreams, religious imagery. The important thing is that you have absolute control here, no matter what you have heard or believed or imagined about death before. If you want a tunnel and white light, you get to have it. If you want to have "choirs of angels sing thee to thy rest," they can be there.

When you are finished, you may want to consider writing another version, giving yourself the opportunity to experiment with a very different sort of story. There are no limits to the way you get to imagine yourself leaving your story, provided that each version be something that you can feel comfortable imagining. The point is not to frighten yourself, but to provide images that give you a sense of control. Whether you write one or several death stories, you can return and revise them or add a new one any time to wish.

The way you leave your storyline is no less within your power to choose than the way you live it. Focusing on one or more versions of your death that you have chosen for yourself allows you to dismiss the fear of any other version and frees you to make life choices unshadowed by the prospect of unexpected death.

Many spiritual masters and mystics, especially from Eastern traditions, have been able to warn their followers and loved ones that they are about to leave their lives. They have given gifts, said their farewells, offered final bits of advice, and then peacefully slipped away. Some have had such control over their own death process that they are said to have settled into a meditative state and simply vacated their bodies, so swiftly and easily that the body was left behind, still seated in the lotus position.

Can the rest of us do such a thing? We can if we believe we can. But even if we aren't yet ready for that level of mastership, we can begin to move ourselves away from a belief that death is completely outside of our control, the one aspect of our lives we cannot influence through the story we tell.

17
Love and Fear

Perhaps everything that frightens us is, in its deepest sense, something helpless that wants our love.
–Rainer Maria Rilke

About fifteen years ago I woke one morning with the thought, *God is love, the Devil is fear.* I lay for a long time contemplating it, wondering where it had come from. This was before I had become acquainted with my author self and learned to recognize this as one of its modes of communication. Though "God is love" was familiar enough (love is one of the "names of God"), the second part was totally new. I did not believe in the God/Devil personae that had been part of my Christian upbringing, but I was very much aware that the good/evil dichotomy (add an o to God to get good, remove the D from Devil to get evil) is fundamental to our experience of reality. The two provide the contrast out of which story is made.

As I thought about the words I had wakened with, I realized that in terms of their effects on our lives, love and fear are opposites. (Never mind that the word my computer thesaurus gives as the antonym for love is hate.) Love (in all its meanings and guises) is the most positive force in our lives, fear the most negative.

What does it do to change God/good to *love* and Devil/evil to *fear*? It moves the locus of control. God and the Devil are cosmic entities—"out there" somewhere, acting on humanity. Good and evil are similarly separate from us, external factors affecting our lives. But love and fear are within us. Love and fear put us in charge of our stories.

If we look deeply into any of our negative emotions, we will find fear at the core. It is fear that *invented* the whole idea of evil, and the image of the devil. Told that God is love or that the Universe is light and goodness (in this case we can think of these as different labels for the same positive force), what could possibly make us doubt such a comforting story? Fear. We'd *like* to accept that we are safe, that we are infinitely loveable and infinitely loved, that our very existence is an exercise in love regarding, exploring and expanding itself—but what if none of that is true? What if the Universe is a cold, lifeless wasteland into which we've been thrown by some capricious force, some random accident? What if love is nothing more than a weak fantasy we've provided for ourselves, a thin blanket to wrap ourselves in when the cold of the lifeless universe becomes too much to bear?

As I've begun talking about Story Principle, I get question after question from people who find it hard to accept, not only because it flies in the face of their previous understanding of Life, the Universe and Everything, but because it feels *too good to be true.* Given the suggestion that it might be possible for someone suffering from cancer to tell a story of healing, a story of aligning with their highest good, so single-mindedly as to create a spontaneous remission, people warn me that even suggesting such a possibility would be giving the patient "false hope." What if someone tries it and it doesn't work? people ask.

When a cancer patient is offered surgery, chemotherapy and/or radiation—all medical interventions *that may or may not work*—doctors are not told to refrain from suggesting them for fear of giving the patient false hope. So it isn't the possibility of failure that creates resistance to putting a positive story to work in our favor. It's the very idea that we have the power of creation within us that seems too good to be true. It doesn't matter how many cases of spontaneous remission one could cite, how many stories of people using their minds and imaginations in favor of healing who were healed by their efforts. The hope these cases provide is called "false" because fear works against our ability to trust such power in our own lives.

Better, we may think, to live as we have before, guarding and guarded, than to open ourselves to the possibility that love and light are everything and only our story about evil, pain and death brings darkness down upon our lives. What if changing to a story of love doesn't work? What if it's not true? What if telling it makes us vulnerable to more of what we are so desperately guarding against?

Lisa, my daughter-in-law, read an early draft of this book and though her reaction was highly favorable in the early chapters, gave me the following response when she finished:

> *If I had been doing my best to put Story Principle to work in my life and my child ran into the street and was killed by a car, I would throw the book against the wall. Story Principle would tell me that my child's death had been my fault—that I was not "good enough" at it to save him! I would reject it just as I reject any religion that blames the bad things that happen in life on lack of faith!*

Her objection helped me to revise and refine the chapter on blaming the victim. And it helped me realize I needed to encourage readers to put Story Principle to work in their lives after the first few chapters (starting with small stories about less critical issues than pain and death), instead of merely reading about it. Until we have experienced the effect of Story Principle, until we have *felt* the way it changes the very context from which we tell our stories and opens us to the effect of love and light in our lives, our old fear story sees it as a threat.

Being the mother of small children, Lisa is at a stage in life when the fear that something might happen to them lurks deep within her consciousness. She is caught between love and fear, the intensity of the love intensifying the fear. So that is why she chose the example she chose. But danger to her sons is only the particular focus of the underlying fear of pain and death that makes our safety in the universe feel like a lie, a false promise.

Fear takes us around and around in a tight circle, keeping itself in the driver's seat. "You're in danger," it tells us. "The moment you let down your guard the enemy will be upon you. You need me to keep you vigilant. Without me you are lost." We are *afraid* to live without fear.

Avoiding danger is no safer in the long run than outright exposure.
Life is either a daring adventure or nothing.
–Helen Keller.

Can fear's promise that it will keep us safe actually protect us? No one gets through life without pain or leaves it without dying. No matter how careful we are, no matter how we limit our own and our loved ones activities, no matter how many precautions we take, fences we build, or security cameras we install, we can't plug every hole through which pain or death might get to us.

Since all of us face pain and death, the quality of our experience of life depends upon the story we tell ourselves about these two apparent enemies. Do we align ourselves with darkness or with light, build our story out of fear or out of love? Fear turns these inevitable experiences into monsters, poised to destroy us, to cast us into eternal darkness. And by focusing our consciousness upon them, fear calls them to us, enlarging them until they shut out the light, fulfilling fear's prophecy.

Love reminds us of what our author self, out beyond story level, knows—that light is all there is, that death and loss are temporary illusions, that pain can be surmounted, healed, transformed. Story Principle helps us tell stories that align us with the light of which we are made. As we practice it, put it to work in our lives, we learn the difference between stories laced with fear and stories founded on love. Our stories become lighter and our well-being inevitably expands. It can do nothing else—because only the resistance we offer to our well-being, resistance built of fear, keeps us caught in the dark.

It was fear that created the scenario of loss Lisa imagined—a loss so painful that she was sure it would lead her to throw away the very tool that could redeem it. Love denies loss, so even if through some hidden faulty wire Lisa's fearful scenario found its way into her experience, telling herself a story of love, of eternal connectedness within a friendly universe would be the surest way to move herself through the darkness step by step back into the light.

Story Principle isn't a protection against pain and death; it's a tool to enable us to see and move beyond them. If surgery, chemo and radiation failed the cancer patient, medicine would have nothing more to offer. If story principle failed to provide a cure, it could still effect a healing.

What allowed Frodo to face his fear and move through his storyline to take the ring to the Cracks of Doom? His love for his threatened world and for the companions on his journey. When Sam, his faithful servant/friend would have killed Frodo's nemesis, Gollum, Frodo's sense of connection and compassion let Gollum live. And it was ultimately Gollum, snatching the ring when Frodo couldn't make himself destroy it, who saved Frodo, his mission and his world. A story aligned with love guides us around the pitfalls, pulls us from the quicksand, defeats the Nazgul and allows us to fulfill the intention of our place in the Saga.

It takes practice to recognize the fear that has worked its way into our stories by insisting that we need it for protection. But as heroes and creators, we get to choose what we include in our stories, and what we keep out, revise, delete.

Imagine Love on one end of a seesaw of consciousness, Fear on the other, and ourselves on the board between. Our fear-drenched culture puts a lot of weight on the Fear end, so most of us have slipped down to join it. The task is to tell ourselves stories that allow us to move up the sloping board toward the middle and beyond, where the "weight" of our consciousness can tip the balance, allowing us to slip easily down to join Love on the other side.

We are all aspects of one consciousness, so each of us who makes that journey up the board, daring to tell a new story focused on well-being, changes the balance point for the whole. As we put Story Principle to use on our own Small Stories, Big Stories, and Foundational Stories, fear doesn't vanish, but its weight diminishes. As we change our stories, bringing more well-being into our own lives, we change the story for all of us, bringing more well-being into our world.

About the Author

Stephanie S. Tolan is the author of 27 books of fiction for children and young adults, one of which (*Surviving the Applewhites*) won a Newbery Honor and another (*Listen!*) the Christopher Award—given to literature that "affirms the highest values of the human spirit." She has also written and co-written plays and musicals, nonfiction articles about the social, emotional and spiritual needs of unusually bright children and adults, and co-authored *Guiding the Gifted Child.* She is a nationally and internationally known speaker and a Senior Fellow at the Institute for Educational Advancement. She lives on a little lake in the woods in Charlotte, NC with her husband, two dogs, a cat and two goldfish. Her websites are www.stephanietolan.com and www.storyhealer.com.

Made in the USA
San Bernardino, CA
06 May 2014